Planning and Control
with PERT/CPM

P9-AOY-655

Planning and Control with PERT/CPM

RICHARD I. LEVIN, PH.D.

CHARLES A. KIRKPATRICK, D.C.S.

*Both of the Graduate School of
Business Administration
University of North Carolina*

McGraw-Hill Book Company

*New York   St. Louis   San Francisco
Toronto   London   Sydney*

Planning and Control with PERT/CPM

Copyright © 1966 by McGraw-Hill, Inc. All Rights Reserved.
Printed in the United States of America. This book, or
parts thereof, may not be reproduced in any form without
permission of the publishers.

Library of Congress Catalog Number 66-21399

5 6 7 8 9 - MUMU - 1

ISBN 07-037365-5

HD
69
P7
L4

G15871

4B

Preface

This is a book about PERT (program evaluation and review technique) and CPM (critical path method). PERT is a method of planning and controlling nonrepetitive projects—jobs that have not been done before and will not be done again in the same precise manner. PERT has contributed to a project as vast as planning the program that lands the first man on the moon; it can contribute to a project as simple as moving a production process from an old plant to a new one. CPM is a planning and control technique used in projects for which some past cost data are available. CPM permits a manager to complete a job in the shortest period of time with minimum expenditures for overtime, additional labor, or additional equipment, and without penalties for finishing late.

The first eight chapters of the book deal with PERT—the ninth with CPM. Much of the material in the PERT chapters is prerequisite to an understanding of CPM. Once a person has grasped PERT, he can acquire an understanding of CPM quickly and with minimum effort.

Here is a brief indication of chapter content:

1. This is the customary and familiar introductory chapter. It presents PERT as an extension of earlier work; it describes what PERT is and does; and it indicates where managers may use PERT.

2. Here the reader is introduced to the fundamentals of PERT —to networks, to events, to activities. He sees that projects can be better planned and controlled when they are broken down into their components.

3. This chapter explains and illustrates a work breakdown schedule. It shows how to get a project ready for the application of PERT and then how to begin to PERT an actual job.

4. The time dimension or variable is introduced here. Simple probability ideas are used in estimating time requirements for projects under conditions of uncertainty.

5. This chapter contains the technical essence of PERT. It shows how to set up networks, how to calculate completion times, and how to monitor and control work.

6. In this chapter the question is asked "How can one reduce the estimated time requirements for a project?" Rearrangement of work, a different sequence for the work, and reallocation of resources are considered.

7. Simple probability procedures are used to determine the chances of finishing the job on time, of finishing before the original completion date, and of finishing before any specified date.

8. The role of the computer in PERT applications is described. The computer makes two major contributions: It can generate a huge amount of information on the status of PERT projects, and it can do this with fantastic speed.

9. This chapter introduces CPM, pointing out that it differs from PERT in that CPM recognizes cost in addition to time. CPM is used to achieve minimum project cost commensurate with the completion date requirement and the cost of expediting the work.

This treatment of PERT and CPM is understandable because it is written in simple terms and uses everyday examples. It carries the reader through those examples step by step, never assuming previous background or competence, anticipating questions and answering them. Probability concepts are included without involvement in complicated statistical theory. Included also is a brief description of twenty-nine current project planning and control techniques that have grown out of the original PERT and CPM techniques. The PERT/CPM bibliography is quite comprehensive.

The authors hope this book will be useful to persons who want to know more about these two recent and challenging tools. They hope it will be a help to managers as they make decisions in today's dynamic environment.

RICHARD I. LEVIN
CHARLES A. KIRKPATRICK

Contents

PREFACE   V

1 Introduction to PERT   1

Repetitive versus nonrepetitive operations   1
Definitions of PERT   2
PERT's background   3
Gantt charts   3
Three transition steps   5
The Polaris project   7
PERT today   8

2 PERT Fundamentals   10

3 Work Breakdown Schedule   19

4 Time Considerations   25

Continuous probability distribution   29
Beta distribution   33

5 Networking Principles   46

Combining networks   46
Zero-time activities   52
Earliest expected date   54
Effects of zero-time activities   60
Critical path   63
Latest allowable date   64
Slack   71

6 Network Replanning and Adjustment   78

Interchanging resources   79
Relaxing the technical specifications   87
Changing the arrangement of activities   88

7 Probability Concepts   93

8 Use of the Computer in PERT Applications   112

When to use a computer   114
Types of computer outputs available   115

9 Critical Path Method   120

APPENDIX **I**   Other Methods of Project Planning and Control 147
APPENDIX **II**   Bibliography of Books, Pamphlets, and Articles 158
APPENDIX **III** Table of Square Roots (1–400)   167
APPENDIX **IV**   Table of Areas under the Curve   171

Index   175

**Planning and Control
with PERT/CPM**

1

Introduction to PERT

The functions of planning, organizing, directing, and controlling are essential to every business regardless of the type, purpose, or complexity of its operation. Techniques, of course, vary because they must be adapted to and appropriate for each individual firm and its own circumstances. PERT, the subject of this book, is a management technique, one which is considerably more useful to some managers than to others. And that leads to a glance at two types of business operations or processes, the *repetitive* type and the *nonrepetitive,* or "once-through," type.

Repetitive versus nonrepetitive operations

Certain business patterns and operations are repeated time after time, month after month, with little or no change. Examples? The manufacture of bricks or steel, of candy bars or towels; the day-to-day operation of a retail store; the periodic and frequent reordering of a standard quantity of operating supplies such as soap or sweeping compound. An excellent example of a repetitive function is production control in a highly automated factory which makes a rather constant quantity each month of a standard product. In repetitive operations and processes, management has experience and data involving standards, quantities, capacities, costs, and time; the need for PERT is slight—PERT's contributions are equally slight.

Nonrepetitive processes differ in that they have not been done before and will not be repeated; for obvious reasons they are referred to as once-through operations or projects. Examples of these? The second canal to connect the Atlantic and the Pacific; advertising sponsorship of a "bowl" game for the first time; a new skyscraper or museum for New York City. Designing a new automobile is essentially a once-through job; making the automobiles on an assembly line is an example of repetitive work. Management must plan, organize, direct, and control nonrepetitive operations; but it is more difficult to do this for nonrepetitive operations than for repetitive operations. Why so? Management cannot be guided by past experience because, by definition, there is no such experience. Imagine the paucity of data and experience available to engineers wanting to build a tunnel under the channel between England and France. Nothing of the sort has been done in anything like identical circumstances before. So, management is always searching for methods and techniques of planning and controlling —of budgeting dollars and time—of directing available resources to attain nonrepetitive goals. PERT is one answer.

Definitions of PERT

PERT stands for Program Evaluation and Review Technique. It was developed by the Navy Special Projects Office in cooperation with Booz, Allen & Hamilton, a management consulting firm.

The PERT technique is a method of minimizing production delays, interruptions, and conflicts; of coordinating and synchronizing the various parts of the overall job; and of expediting the completion of projects. It permits the turning out of work that is controlled and orderly. It is a method of scheduling and budgeting resources so as to accomplish a predetermined job on schedule. It is a communication facility in that it can report developments, both favorable and unfavorable, to managers and in that it can *keep* the managers posted and informed. Above all, PERT is an outstanding approach to achieving *completion of projects on time*.

One brief word of caution: Neither PERT nor any other management technique solves managers' problems. Instead, they help a manager realize what his problems are, what solutions are realistic, and hopefully, the strengths and weaknesses of each. PERT tries to keep managers apprised of all factors and considerations that bear on their decisions. Although PERT is no substitute for managerial intelligence and perceptivity, experience and judgment, it can be a most worthwhile aid and tool in decision making.

PERT's background

GANTT CHARTS

PERT did not appear as a new and revolutionary tool of management. As have many of our management techniques, it grew from earlier attempts by managers to gain better control over what they had to manage.

Most students of management are familiar with the name H. L. Gantt. Gantt was a contemporary of "the father of scientific management," F. W. Taylor. In his work on production control, Gantt developed the now-famous Gantt chart still found today in many production control offices; indeed, some of his advanced ideas may well have been the forerunners of PERT.

Gantt used what he called a "Gantt milestone chart"; this was basically a chart depicting work to be done, but what is more important, it also denoted the interrelationships between and among all phases of this work. In other words, the chart showed in a modest manner the coordination required between the various phases of each of the projects.

Figure 1–1 illustrates one of Gantt's milestone charts. By using such a Gantt milestone chart, with its time scale across the bottom, one can easily see how long it will take to accomplish a specific project. Each of the circles (milestones) represents the

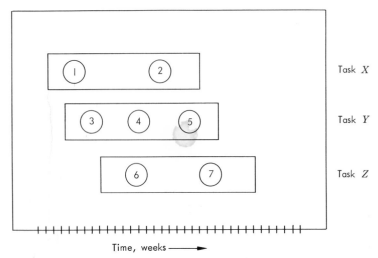

Figure 1-1.   Gantt milestone chart.

accomplishment of a specific phase of the total undertaking, and of course, each of the rectangles represents a task. The three rectangles taken together represent the entire project.

The Gantt chart showed the relationship between the milestones within the same task; one can see from Figure 1–1 that milestone 2 cannot be started until milestone 1 has been completed. One can also see from looking at the three milestones in task Y that milestone 4 cannot be begun until milestone 3 has been completed. But what about the relationship between task X and task Y, or between task Y and task Z? The Gantt chart does not tell us whether milestone 6 can be begun before milestone 2 has been completed, or indeed, whether milestones 6 and 7 are dependent at all upon the milestones in task X. Therein lies the big limitation of this particular tool. In short, we can tell from a Gantt milestone chart the relationship between two specific milestones within a task, but the relationship between and among milestones contained in different tasks is not indicated on the chart.

The developers of PERT improved on Gantt's original milestone chart and modified it in order to illustrate interrelationships between and among all the milestones in an entire project. This was accomplished basically in *three* steps. Following them through closely and in detail will enable us to appreciate more fully some of the advantages of PERT over the Gantt planning and control technique.

THREE TRANSITION STEPS

Figure 1–2 indicates the results of the *first* step in the transition from Gantt's chart to a PERT network. Essentially what was done in this first transition step was to remove the rectangles signifying tasks and to illustrate the interrelationships between milestones within a specific task by the use of arrows connecting the milestones. Although this resulted in a somewhat neater chart, still unsolved was the problem of interrelationships between milestones not contained in the same task. For example, is it necessary for milestone 4 to be completed before milestone 7 can begin? Can the two milestones contained in task *X* proceed independently of the three milestones contained in task *Y*—or must milestone 3 be completed before milestone 2 can start?

Gantt's chart was an effective scheduling tool with respect to simple, uncomplicated projects involving a minimum of coordination among the various tasks making up the entire project. But when one considers using this particular technique to represent the enormous number of interrelationships among the phases of the

Figure 1–2. Results of the first step in transition from Gantt milestone chart to PERT network.

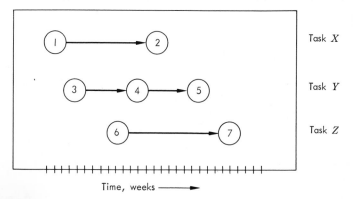

Time, weeks ⟶

development of a Polaris project, treated shortly, he can see how inadequate it would be as an effective planning and control tool.

The *second* transition step between the milestone chart and the PERT network involved adding the relationships among milestones contained in different tasks, as in Figure 1–3. The addition of arrows representing the necessary flow of the work now clears up our previous questions about what milestones must precede what other milestones, and which milestone can be completed without being coordinated with which others. For instance, we can easily see that milestone 4 cannot be begun before milestone 3 has been completed, and because milestone 3 depends on milestone 1, we can say that milestone 4 also depends on milestone 1. In short, the relationships between and among all milestones in the entire project (regardless of which task the milestone is a unit of) are clearly shown on this chart by use of the added arrows. Milestone 1, of course, is the beginning point of the entire project, and milestone 7 is clearly the ending point of the entire project.

Now that Figure 1–3 indicates all the interrelationships between and among all the milestones, of what use are the task designations? None—all relationships regardless of the task involved are clearly shown by the arrows. For this reason, the *third* transition step omits the term "task" entirely; this has been done

Figure 1–3. Results of the second step in transition from Gantt milestone chart to PERT network.

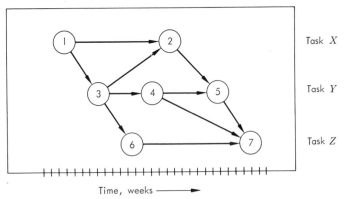

Time, weeks ———▶

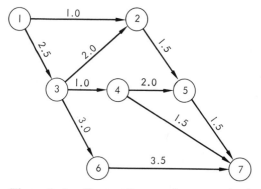

Figure 1–4. Gantt milestone chart completely transformed into PERT network.

in Figure 1–4. In addition, the horizontal time scale is omitted in favor of individual time notations on the arrows themselves. Because all times are given, a time scale is unnecessary. The completed transaction is shown in Figure 1–4 with some arbitrary times in weeks indicated over the arrows.

Thus, the Gantt milestone chart, which showed milestones or achievements within defined tasks and linked tasks to form a finished project, gave way to a PERT network. The major advantages of the network over the milestone chart are that the PERT network:

1. Indicates all the interrelationships among all milestones
2. Makes redundant the designation of the task because the project is viewed as an integrated whole and not as a number of tasks
3. Substitutes for the previous time scale an individual time value for each leg of the network
4. Allows the use of the network for highly complicated projects
5. Allows the use of probability theory for estimation of completion dates when there is uncertainty

THE POLARIS PROJECT

As we mentioned when we described what PERT is and does, the Navy Special Projects Office was prominent in the development of PERT and, in addition, is credited with its first major use. In the 1950s, when the government was exploring the possibilities and the potentialities of nuclear-powered submarines, the *technical* and *engineering* problems associated with such a new project were assumed to be the most challenging problems. However, one of the top officers working on the program suspected that the really significant problems, the really difficult ones, were not technical but rather were the problems of coordinating such a vast undertaking. In short, he realized that the overall management problem, specifically the planning, coordinating, and control of all the resources required, would dwarf the technical problems.

To give some idea of the vastness of the problem, a method had to be found by which those in charge of the Polaris project could deal with 250 prime contractors and at least 9,000 subcontractors. These numbers may not at first appear to represent such a staggering task until one realizes that the failure of any one of the subcontractors to deliver one small, seemingly insignificant part of hardware might slow up or even stall the entire project. Then when one considers the enormous number of possibilities for this type of delay, the problem assumes a magnitude which, according to authorities, overshadowed even the Manhattan project, the development of the first atomic weapon.

A method had to be developed which would allow the management of the Polaris project to coordinate the efforts of all these firms, to anticipate the occurrence of bottlenecks, to forecast with reasonable certainty the extent to which target dates would be met, and in general to channel the efforts of literally hundreds of thousands of persons into a finished operational weapon. PERT was the solution. The Navy Department has stated on several occasions that because of the use of PERT, the Polaris missile submarine was brought to combat readiness about two full years ahead of the original date scheduled.

PERT today

No longer is PERT mainly identified with and used by the military. Manufacturers, builders, and advertising agencies are types of commercial firms that use the tool. Where? In research and development, in construction, in the launching of new products, and in various marketing activities. Recently a stadium was built in Atlanta for major-league baseball and football. Construction time was held to less than twelve months through the use of the critical path method (CPM), a variant of PERT. Chapter 9 takes up CPM.

The authors are acquainted with a contracting company which uses PERT on every one of its construction projects involving $150,000 or more. One of the principals of this company recently stated that although PERT was not a panacea for all his problems, it did represent an orderly way of thinking about a construction project and of seeing far in advance trouble spots which would soon represent real problems. When questioned on the effectiveness of this technique, he estimated that the use of PERT as a planning, coordinating, and control technique by his company had in general reduced the time required to perform a given type of work by an average of about 10 percent. Readers who are familiar with the fixed costs associated with a construction project can well appreciate the tremendous savings this company has realized from the use of PERT in its work.

Today PERT is both forcing and permitting managers to think each major program and project through in its entirety and in detail. It identifies possible delays for managers, and it aids in resolving the difficulties. It helps managers make earlier deadlines because of continuous, effective control. PERT contributes to the optimum utilization of resources, particularly money, time, and manpower.

2

PERT Fundamentals

PERT is concerned with two concepts:

1. **Events:** An event is a specific accomplishment that occurs at a recognizable point in time.
2. **Activities:** An activity is the work required to complete a specific event.

In PERT networks, events are generally drawn as circles, and activities are represented as arrows joining two circles. Figure 2–1 illustrates two events connected by one activity. The events have been assigned numbers so that we can identify them. Each of the two events represents a specific point in time; event 1 could represent the point in time "work started," and event 2 could represent the point in time "work finished." The arrow or activity connecting these two events would then represent the actual work done; it represents *time*—the time needed to plan and do the actual work.

Figure 2–1. Two events connected by one activity.

We can notice from this simple illustration that events take no time in themselves but only mark the beginning or ending of an activity. Thus, in PERT the activities, not the events, require time, money, and resources. In this sense H. L. Gantt was quite right in referring to his events as milestones, the implication being that they denote the beginning or the completion of work, but never the work itself.

At this point two formal definitions are in order. An event in PERT is:

> An accomplishment occurring at an instantaneous point in time but requiring no time or resources itself.

An activity is:

> A recognizable part of a work project that requires time and resources for its completion.

Suppose digging a basement is to be the first step in building a house. This first step is expressed in PERT terminology in Figure 2–2. Event 1 would be described as "basement excavation begun" or "basement excavation started"; event 2 would be described as "basement excavation finished" or "basement excavation completed." The activity, which incidentally would be called in the PERT system activity 1-2 (indicating its beginning and ending events), would be described as "excavate basement" or "dig basement."

It can easily be seen from this simple example that events 1 and 2 do not in themselves require any time; they merely signify the beginning and the end of work. Activity 1-2 denotes the actual work required. Notice in the PERT network shown in Figure 2–2 that the activity flows from left to right; this is the general rule

Figure 2–2. Repeat of Figure 2–1.

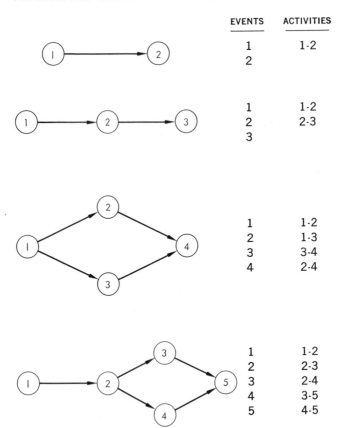

	EVENTS	ACTIVITIES
	1	1-2
	2	
	1	1-2
	2	2-3
	3	
	1	1-2
	2	1-3
	3	3-4
	4	2-4
	1	1-2
	2	2-3
	3	2-4
	4	3-5
	5	4-5

Figure 2–3.   Simple networks described in proper PERT terminology.

for all PERT networks. Also notice that between two events there can be only one activity, and that this activity is defined by its beginning and ending events.

Figure 2–3 shows several very simple PERT networks with their events and activities described in proper PERT terminology. Thus, using the conventions shown in Figure 2–3, we can distinguish between any events, and we can refer to any specific activity within the network. This relationship between events and activities is the basic notion which underlies PERT.

Table 2–1. Description of Activities in Constructing a Pipeline

ACTIVITY DESCRIPTION	DESIGNATION	BEGINNING EVENT	ENDING EVENT
Lay out pipeline on ground	1-2	1	2
Dig trench	2-3	2	3
Lay joints of pipe in trench	3-4	3	4
Connect pipe	4-5	4	5
Cover pipe with dirt	5-6	5	6
Tamp down earth	6-7	6	7
Inspect work	7-8	7	8

The term *network* denotes that when several events and activities are combined and the resulting diagram is drawn, that diagram takes on the general appearance or shape of a network. There may be several branches, of course, depending upon the complexity of the project represented.

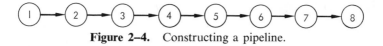

Figure 2–4. Constructing a pipeline.

For instance, the network represented in Table 2–1 and Figure 2–4 indicates the events and activities required to construct a small pipeline between a well and a summer house. First each of the activities has been described in Table 2–1, together with its beginning and its ending event, and then all the events and activities have been placed together to form the resulting network in Figure 2–4.

The network described by Figure 2–4 is obviously a straight-line type of network; that is, each of the activities requires that the immediately preceding activity be completed before the next activity can begin. Many networks, of course, are not of this type. If, for instance, we wanted to illustrate the construction of two identical pipelines and we wanted to indicate specifically that the work on one of these could go on independently of the work being done on the other, our resulting network would take the shape of that described in Figure 2–5. We have assumed here that each of

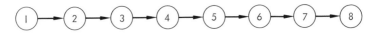

Figure 2–5. Constructing two independent pipelines.

the pipelines contains the same seven activities described in Figure 2–4.

If we had wanted to regard the laying of the two pipelines as one integrated project, we could have added an artificial beginning event and an artificial ending event as has been done in Figure 2–6. By adding events 17 and 18, we have implied through the use of *one* network that the laying of the two pipelines is an integrated project. Event 17 would be described as "begin project" and of course event 18 as "project completed." We must remember that activities 17-9, 17-1, 8-18, and 16-18 are artificial in the sense that they do not require any time or resources, and are put in simply to show the two pipelines as a single project. Because of their nature, these types of activities are commonly called *zero-time activities*.

In Figure 2–6 the activities denoting the work required to complete pipeline 1 (activities 1-2, 2-3, 3-4, 4-5, 5-6, 6-7, and 7-8) could proceed independently of the activities required to complete the other pipeline. For example, activity 6-7 could go on before activity 10-11, and so on. However, within one of the pipelines, the top one, for instance, activity 5-6 could not be begun before activity 4-5 had been completed.

Figure 2–6. Constructing two pipelines as an integrated project.

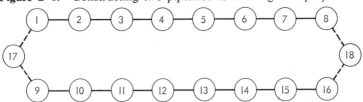

The activities on the top pipeline, 1-2, 2-3, 3-4, etc., are commonly referred to as being *series-connected*; that is, no activity can be begun before the immediately preceding activity has been completed. If, however, we want to describe separated activities (4-5 and 9-10, for example), we say that they are in *series-parallel*; this means that each can proceed independently of the other. We obviously could not consider the combined project (two pipelines) completed until all the activities on both pipelines had been completed.

Most PERT networks do not come out as parallel lines or as simple series-connected events. When there are complex interrelationships within a network, of course, they must be illustrated by the network diagram. Examine the network illustrated in Figure 2–7. From our previous contact with network fundamentals we know that event 1 is the first event in the network. It is commonly called the *network-beginning event* and is so recognized because no activities lead to it. Similarly, we see that event 8 is the *network-ending event*. It is recognized as such because no activities lead away from it.

We can also pick out some other relationships from Figure 2–7. Looking for a moment at event 6, we notice that activities 5–6 and 2–6 lead to it; therefore, we can say that event 6 is the ending event for two activities. Using the same line of reasoning, we notice that event 2 begins two activities, activities 2–5 and 2–6; thus, we can state that event 2 is the beginning event for

Figure 2–7. Relationships among events.

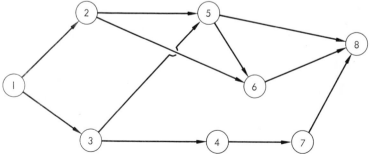

Table 2–2. Review of Network Fundamentals

ACTIVITIES	BEGINNING EVENT	ENDING EVENT
1-2	1	2
1-3	1	3
2-5	2	5
2-6	2	6
3-5	3	5
5-6	5	6
3-4	3	4
4-7	4	7
5-8	5	8
6-8	6	8
7-8	7	8

ENDING EVENTS	ACTIVITIES FOR WHICH IT IS AN ENDING EVENT
2	1-2
3	1-3
4	3-4
5	2-5, 3-5
6	2-6, 5-6
7	4-7
8	5-8, 6-8, 7-8

BEGINNING EVENTS	ACTIVITIES FOR WHICH IT IS A BEGINNING EVENT
1	1-2, 1-3
2	2-5, 2-6
3	3-5, 3-4
4	4-7
5	5-8, 5-6
6	6-8
7	7-8

NETWORK-BEGINNING EVENT

1

NETWORK-ENDING EVENT

8

two activities. Event 8 is the ending event for three activities as well as being the network-ending event.

As a review of network fundamentals, Table 2–2 lists all the activities contained in Figure 2–7, together with their beginning and ending events; in addition, all events which are beginning events for some activity and all events which are ending events for some activity are identified. Finally, the network-beginning and network-ending events are listed.

The network in Figure 2–7 and Table 2–2 demonstrates that a single event can be both a beginning event for an activity and an ending event for another activity. Take, for instance, event 5 in Figure 2–7. This event is a beginning event for activities 5–8 and 5–6, and at the same time it is an ending event for activities 2–5 and 3–5. In similar manner, a single event can be the ending event for an entire group of activities and at the same time the beginning event for still another group of different activities. As an illustration, observe event 4 in Figure 2–8. This event is the ending event for a group of activities (2–4 and 3–4) and at the same time it is the beginning event for another group of activities (4–5 and 4–6). It is not uncommon for events to play this dual role in many PERT networks.

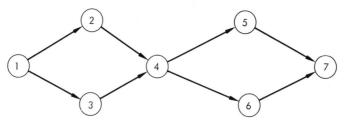

Figure 2–8. Network in which one event is the ending event for one group of activities and the beginning event for another.

As a review exercise, we have listed in Table 2–3 several activities with their beginning and ending events. Using this information, see if you can draw the proper network which should result from these interrelationships. The properly drawn network is shown in Figure 2–9.

Table 2–3. Several Activities with Beginning and Ending Event

ACTIVITY	BEGINNING EVENT	ENDING EVENT
1-2	1	2
1-3	1	3
3-4	3	4
2-4	2	4
1-4	1	4
4-5	4	5
4-6	4	6
4-7	4	7
5-7	5	7
6-7	6	7
2-5	2	5
3-6	3	6

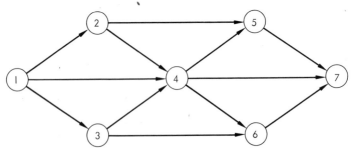

Figure 2–9.   Properly drawn network for the activities in Table 2–3.

3

Work Breakdown
Schedule

Thus far, we have looked at the basic idea of PERT; we have 'defined and illustrated the differences between events and activities, the fundamental elements of the PERT network. Now we must examine briefly the preliminary work that is necessary before the actual construction of a PERT network can begin.

As in other management endeavors, the first step in the development of a PERT network is the establishment of objectives: the major objective to be accomplished and each of the supporting objectives in the entire project. When the major objective and each of its supporting objectives have been identified, they must be linked together so that the planner can see the project in its true perspective, so that he can see the relationships between and among all the steps.

In simple work projects, the major objective and each of the supporting objectives are so easily and quickly seen that complicated listings or arrangements are not required. For instance, in the building of a small house (which usually consists of about seventeen steps, beginning with site clearing and ending with landscaping) the typical contractor can easily visualize the major objective, the completion of the house on time. Through experience with houses of similar design, he can easily see the interrelationships among the seventeen steps required for completion of the house.

In more complex projects, however, such as the construction of a shopping center complex with 100 stores, the thousands of individual events and activities are not so easily recognized and related as they would be in the simple house-building example just mentioned. For this reason, some users of PERT make use of a *work breakdown schedule,* a pictorial representation of the entire program. Now, this schedule is *not* a PERT network; instead, it is a preliminary diagram of the way in which all the supporting objectives go together and mesh to ensure the attainment of the major objective.

On a very complex project consisting of many thousands of events and activities, the work breakdown schedule aids in the identification of objectives; it allows the planner(s) to see the total picture of the project, though not in completely detailed form; it provides a look at the forest but not at the individual trees. For example, let us construct a simple work breakdown schedule for building a house. Admittedly, such a schedule would often be unnecessary because the typical builder is familiar with the essential steps in such a project. However, as most people who are not builders are at least somewhat familiar with the basic stages of house construction, this project will be a good example for us.

The work breakdown schedule begins with the major or total objective, in this case "house constructed." This goes on the top or major level of the work breakdown schedule. As in an organization chart, the second level will contain the supporting objectives or stages in the house construction. Then the third level will indicate some of the more detailed work required to complete these second-level supporting objectives until the entire construction of

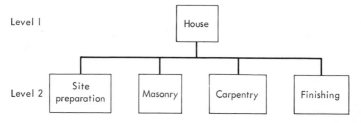

Figure 3–1. Work breakdown schedule for building a house (levels 1 and 2).

the house has been plotted somewhat in the form of an organization chart. Figure 3–1 illustrates the major objective on level 1 and the supporting objectives on level 2. This indicates that in order to attain the major objective (construction of the house), four subobjectives must be attained: site preparation, masonry, carpentry, and finishing. To continue the expansion of the work breakdown schedule, we would of course add level 3.

In this illustrative case, since the addition of the *entire* level 3 would require too much space, we have elected to illustrate the development of the work breakdown schedule by concentrating upon the masonry phase of the project. The schedule with level 3 masonry added appears in Figure 3–2. Through use of a work breakdown schedule we have now shown that there are four sub-

Figure 3–2. Work breakdown schedule for building a house (levels 1, 2, and 3).

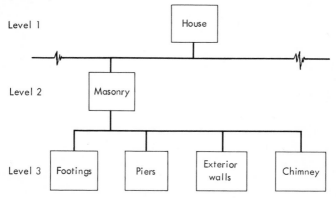

objectives on level 2; for one of these subobjectives (masonry) there are four distinct phases: preparation of the footings (supports for the walls), preparation of the piers (supports for the interior of the house), laying of the exterior walls, and construction of the chimney.

Level 4 of the house project is illustrated in Figure 3–3. Again for purposes of brevity and space allocation, we have limited this level to expansion of the footings phase.

No one can state dogmatically or precisely just how detailed the work breakdown schedule should be or exactly how many levels are needed for any one project. The basic demand is that the work breakdown schedule be detailed enough to allow the eventual construction of a PERT network which will accurately reflect the interrelationships among all the events and activities which make up the entire project. In our house-building example, certainly four levels would suffice. In a large, complex project such as the development of a new automobile model, however, as many as ten or even more levels of subobjectives might be required.

Figure 3–3. Work breakdown schedule for building a house (levels 1, 2, 3, and 4).

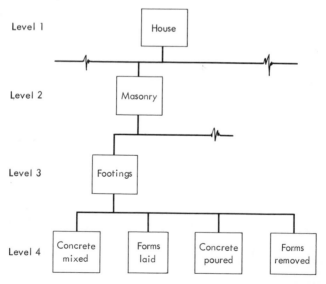

In any event, the more levels on our work breakdown schedule, the more detail we are using.

Notice that the number of boxes or entries on any one level of the work breakdown schedule is not the same as the number on another level of the same schedule; each number depends entirely upon the relative complexities of the suboperations on that level. From our example of drafting the work breakdown schedule for building a house, you can quickly see that PERT is essentially a "from-the-top" type of planning. That is, the major objective is established first, then it is broken down into its subobjectives, then each subobjective is broken down into *its* subobjectives, and so on, until the entire project is reduced to a degree of detail which makes it manageable. Obviously, to look at a very complex project and to attempt to see the interconnections among all the events and activities required would be impossible without first making this breakdown.

Our conclusion with respect to the use of a work breakdown schedule? It should usually be prepared and used prior to the construction of a PERT network on every large or complex project. Its major benefit? It presents the entire project in a systematic way so that the interrelationships among all phases of the project are easily seen.

At the start of the preparation of the work breakdown schedule, the various individuals who will be concerned with the management of the project, from both a technical and a managerial point of view, will have to begin working together. For optimum results in the utilization of PERT, both a technical representative and someone knowledgeable in PERT should participate in this phase of the work. A person familiar with construction of a house, for instance, would not usually be familiar with applications of the PERT techniques; by the same token, a specialist in the uses of PERT would not normally be familiar with the technical details of many of the projects on which he would need to apply PERT. Each type of person, then, is usually needed: (1) an expert in the technical phases of the project work to be controlled and (2) an expert in the techniques of PERT.

Each person who participates in the application of PERT to the control of the project should have some basic familiarity with the general nature of the work and with the ultimate objective desired. This does not require each team participant to be an overall work expert if the project is, for example, the design of a missile. It does imply, however, that all the members of the PERT team should be familiar enough with the general problems involved in missile development so that they can at least communicate intelligently with the other members of the project team.

4

Time Considerations

Because PERT has to do mainly with the control
of projects, we obviously must come shortly to the
consideration of *time*. Time becomes in PERT the
basic measure of how long our project will take, how
much time ahead of or behind schedule we are at any
point, and what work content is in each phase or
activity of our project. Time is clearly a most essential and basic variable in the PERT system of planning and control. There are considerations in PERT
other than time, but time is the basic measure of the
work required in a project.

In PERT, time is usually expressed in calendar
weeks. The basic reason for not using units of smaller
magnitude is that most of the activities in a PERT
network will take considerable time to accomplish.
Thus, if we used 1 hour as our basic measure and
one of our activities required 7 weeks for com-

pletion, the units would become somewhat large for easy handling:

7 weeks × 40 working hours / week = 280 working hours

When we use weeks, on the other hand, the above time requirement is simply expressed as 7 weeks. For another example, suppose again that hours were used instead of weeks and further suppose that a project expected to require 10 weeks (10 × 40 = 400 hours) actually consumed 401 hours. To express the extent to which the estimated time had been exceeded, we would use the calculations 401 − 400 = 1 hour. To express this 1-hour overage as a percentage of the total estimated time requires that we compute 1/400, or about .25 percent. Now, to a practical project manager, the significance of 1 hour out of 400 is questionable. Since we are dealing with estimates in the first place, using 1 hour as the basic unit of measurement on a project of some months' duration would perhaps represent spurious accuracy. This is one reason why the basic time unit in PERT has been established as 1 week.

To express the time required to complete an activity in the PERT operation, we first estimate the number of working days required to do this work and then divide this by the number of working days per week. For example, if we have an activity that is expected to require 10 working days and our normal working week is 5 days, then we would say in the PERT system that this activity is expected to require 2 calendar weeks. As a formula this calculation would be expressed as follows:

$$\text{Time in calendar weeks} = \frac{\text{number of working days required}}{\text{number of working days per week}}$$

The quotient is usually expressed to one decimal place. That is, we find such answers as 2.3 weeks, 5.6 weeks, and 6.8 weeks; rarely do we see such answers as 2.34 weeks, 5.64 weeks, or 6.81 weeks. There is good reason for this. The times required for completion of the various activities in a network are all products of estimates; to carry out the answers to more than one decimal place would create an artificial degree of accuracy which is neither correct mathematically nor helpful to a project manager.

For illustrative purposes, we have listed in Table 4–1 several

Table 4–1. Computation of Time Estimates

NUMBER OF WORKING DAYS REQUIRED	NUMBER OF WORKING DAYS PER WEEK	NUMBER OF CALENDAR WEEKS REQUIRED
40	5	8.0
42	5	8.4
43	5	8.6
16	6	2.7
19	6	3.2
23	4	5.8
26	4	6.0
13	7	1.9
16	7	2.3
14	7	2.0
81	5	16.2
122	6	20.4

estimates of the number of working days required to complete certain activities together with the typical number of working days per week; we have also computed the time required, using the rules laid down above. We have followed the normal rules of mathematics about rounding off; any digit 5 or greater in the second place to the right of the decimal raises the digit in the first place by 1; any digit less than 5 is rounded to 0.

You recall that the major use of PERT is as a planning and control tool on projects which are referred to as once-through. This means that the persons charged with the responsibility of estimating how much time each activity in the network will take have had little or no experience with exactly this type of work in exactly the same form. Therefore the time budgeted for the completion of each of the activities in the network must unavoidably be an estimate. When one says that activity 2-3, for instance, is expected to take 10.0 weeks, he is really saying that he *thinks* it will take 10.0 weeks. And he knows full well that because of weather, contractors' failure to deliver materials, machinery breakdowns, and such, the activity may really take 12.0 weeks, or even in extreme cases 20.0 weeks.

On the other hand, the estimator is aware of the possibility that the activity may be completed in less than the estimated time.

Assume, for instance, an activity involving the digging of drainage ditches outdoors. An estimator knows that the chances are small of having 20 days uninterrupted by rain, so he may well include in his time estimate an allowance of several days for interruptions by rain. Suppose his estimate, including this built-in allowance for inclement weather, comes out to be 4.5 weeks. Now suppose that work commences and there are 4 consecutive weeks of sunny weather. It is quite likely that the work will be done in less than 4.5 weeks.

Chance can permit an activity to be completed in less than the estimated time. Suppose that an activity in a particular network involves the drilling of a test shaft underground which must connect with another shaft at a distance of some feet. Suppose further that past experience with this type of work indicates that the chances are 1 in 5 of getting the shaft drilled correctly the first time. A person charged with estimating the time required for this activity will doubtless take this into account in his estimate. He will allow extra time because the odds are only 1 to 5. Assume that he does allow this safety factor and that his estimate of the time for this activity is 2.5 weeks. The drilling crew goes to work and, as luck would have it, does a perfect job the first try. It is then obvious that the time required for completing this activity will be considerably less than that estimated in the beginning. Because much of the work involved in projects of this type has not been done before, some uncertainty is involved in estimating how long each activity will take. The uncertainty in estimating completion time for each of the activities does of course interject uncertainty into the estimate of the time required to complete the entire network.

These facts did not go unnoticed by the developers of PERT; those men were fully cognizant of the uncertainty involved in estimating times for jobs which had not been done before. They therefore developed methods for coping with this uncertainty. Notice that we say *coping* with it. There is obviously no way to remove uncertainty; this is simply a fact of business life in general. There are, however, methods by which we can know just how certain we are and just how much reliance we can put on a figure representing an estimate that someone gives us. The methods used to cope with uncertainty involve some knowledge of statistics and more particularly of probability theory. As some of our readers will not have had the opportunity to work much with probability theory, we have included a brief introduction to the subject. By no means is this a complete treatment of the theory of probability; instead, it is a simple background or foundation that will enable you to understand the fundamentals of coping with uncertainty. The authors felt that instead of providing a few rote methods or unexplained rules, they should take a little time and present an introduction to probability theory.

Continuous probability distribution

We have just recognized that the time required to do a certain job or a particular piece of work can vary. Many factors or influences can function as determinants of the length of this time period. When everything works perfectly, elapsed time is delightfully and unexpectedly short; when everything goes wrong, the job seems interminable. Somewhere between the extremes is the concept of "average" or typical time.

Observation of many phenomena in our daily experience reveals the same type of range of magnitudes. There are, for instance, a very few families whose annual income is less than $1,000 *and* there are a very few families with annual incomes of over $1 million. Most families have incomes that fall somewhere

Table 4–2. Daily Sales of a Certain Product

QUANTITY SOLD		
26	18	13
13	9	19
33	9	18
10	10	10
13	17	17
5	20	13
7	19	22
10	5	9
3	18	7
13	17	17

between $5,000 and $15,000. When referring to this range, we may note that most incomes are *distributed* or spread out between the two extreme figures, $1,000 and $1 million. When we spot or locate many incomes between the two extremes, we form a distribution known as a *continuous* distribution.

Consider, for example, a case in which 30 daily sales of a certain product are as listed in Table 4–2. We then take these values for past sales and plot them on a graph as Figure 4–1. When we draw a rough line through the points, we find that this line takes the *approximate* shape of the often-referred-to "bell-shaped" curve. A more nearly correct name for this would be the "normal" curve. We can, of course, calculate the average sales simply by dividing the total quantity sold during the 30-day period by 30 as follows:

$$\text{Average sales per day} = \frac{420}{30} = 14$$

Exactly what is the significance of the normal curve? Simply this: In most groups of historical data there are a few values which

Figure 4–1. Continuous distribution of past daily sales.

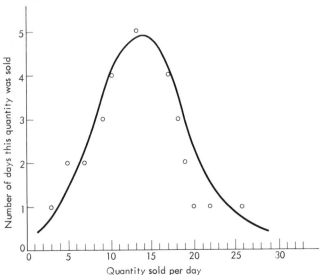

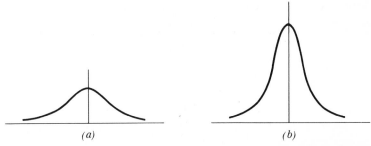

(a)   *(b)*

Figure 4–2. Two varient curves. (*a*) Values are widely dispersed away from their average. (*b*) Values are tightly grouped about their average.

are exceedingly small and a few values which are quite large; but most of the values tend to group or cluster around the average in much the same manner as our graph of past sales indicates. For instance, there are undoubtedly some individuals who are less than 3 feet tall and, of course, there are some who are over 8 feet in height, but most are between 5 and 6 feet tall. Important for our purposes is the fact that most groups of data such as these tend to take the bell shape when plotted. We shall assume that the actual values of past daily sales are normally distributed around their average of 14. We make this assumption because the bell-shaped distribution is in many cases a reasonable approximation of business phenomena.

In addition to this curve, there are curves in which the values tend to be widely dispersed away from their average and other curves in which the values tend to group themselves tightly around their average. These two curves are illustrated in Figure 4–2.

There is a statistical measure of the tendency for data to disperse around their own average. This measure is called the "standard deviation." Because we can make important management inferences from our past data with this measure, it is useful to learn how to calculate it and what it means.

The standard deviation is calculated by following these five steps:

1. Subtract the average from each value in the data.
2. Square each of the differences obtained in step 1.
3. Add together all the squared differences.

Table 4–3. Computation of the Standard Deviation for the Data in Table 4–2

1. SUBTRACT THE AVERAGE FROM EACH VALUE	2. SQUARE EACH OF THE DIFFERENCES	3. ADD THE SQUARED DIFFERENCES
$26 - 14 = 12$	$(12)^2 = 144$	144
$13 - 14 = -1$	$(-1)^2 = 1$	1
$33 - 14 = 19$	$(19)^2 = 361$	361
$10 - 14 = -4$	$(-4)^2 = 16$	16
$13 - 14 = -1$	$(-1)^2 = 1$	1
$5 - 14 = -9$	$(-9)^2 = 81$	81
$7 - 14 = -7$	$(-7)^2 = 49$	49
$10 - 14 = -4$	$(-4)^2 = 16$	16
$3 - 14 = -11$	$(-11)^2 = 121$	121
$13 - 14 = -1$	$(-1)^2 = 1$	1
$18 - 14 = 4$	$(4)^2 = 16$	16
$9 - 14 = -5$	$(-5)^2 = 25$	25
$9 - 14 = -5$	$(-5)^2 = 25$	25
$10 - 14 = -4$	$(-4)^2 = 16$	16
$17 - 14 = 3$	$(3)^2 = 9$	9
$20 - 14 = 6$	$(6)^2 = 36$	36
$19 - 14 = 5$	$(5)^2 = 25$	25
$5 - 14 = -9$	$(-9)^2 = 81$	81
$18 - 14 = 4$	$(4)^2 = 16$	16
$17 - 14 = 3$	$(3)^2 = 9$	9
$13 - 14 = -1$	$(-1)^2 = 1$	1
$19 - 14 = 5$	$(5)^2 = 25$	25
$18 - 14 = 4$	$(4)^2 = 16$	16
$10 - 14 = -4$	$(-4)^2 = 16$	16
$17 - 14 = 3$	$(3)^2 = 9$	9
$13 - 14 = -1$	$(-1)^2 = 1$	1
$22 - 14 = 8$	$(8)^2 = 64$	64
$9 - 14 = -5$	$(-5)^2 = 25$	25
$7 - 14 = -7$	$(-7)^2 = 49$	49
$17 - 14 = 3$	$(3)^2 = 9$	9
		1,264

4. DIVIDE THE SUM OF THE SQUARED DIFFERENCES BY THE NUMBER OF VALUES

$$\frac{1,264}{30} = 42.13$$

5. TAKE THE SQUARE ROOT OF THE QUOTIENT FROM STEP 4

$$\sqrt{42.13} = 6.49$$

4. Divide the sum of all the squared differences by the number of values.
5. Take the square root of the quotient obtained in step 4.

In Table 4–3 these five operations have been performed on the original data from Table 4–2. The standard deviation of the distribution of past daily sales is 6.49 units. Now, what does this mean? Just this: There is mathematical proof that:

1. Approximately 68 percent of all the values in a bell-shaped distribution lie within ±1 standard deviation from the average.
2. About 95 percent of all the values lie within ±2 standard deviations from the average.
3. Over 99 percent of all the values lie within ±3 standard deviations from the average.

Now let us apply these to our data.

If the average of our past daily sales is 14 and if our curve is perfectly bell-shaped, then approximately 68 percent of all future sales will fall between 14 + 6.49 units and 14 − 6.49 units— between 20.49 units and 7.51 units. Similarly, about 95 percent of all future sales will fall between 14 + (2 × 6.49) units and 14 − (2 × 6.49) units—between 26.98 and 1 unit. There are statistical tables available which will indicate that portion of all the values in a distribution which is contained within *any* number of standard deviations from the average. These tables will be used in Chapter 7 to help determine the chances of completing a project on time. We shall defer additional discussion of these statistical tables until that time.

Beta distribution

Not all variables are normally distributed; not all take the form of a bell-shaped curve. Let us begin this section, however, with a normally distributed variable, one which does take the familiar bell shape. Figure 4–3 is a normal curve for height of adult males. Exactly what does Figure 4–3 indicate? Just this: The average height of males is about 5 feet, 9 inches; there are a few adult males 4 feet, 6 inches, and a few males 7 feet tall. As you can easily see, most of the adult males tend to cluster around the average of 5 feet, 9 inches.

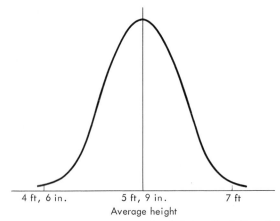

Figure 4–3.   Normal curve for height of adult males.

If we repeat Figure 4–3 and add to it, we can illustrate a basic feature of the probability distribution which will become important to us in the PERT analysis. Figure 4–4 illustrates the same distribution, the height of adult males, but we have added three lines, *A, B,* and *C.* In probability theory each of these lines

Figure 4–4.   Normal curve for height of adult males, with three probability values shown.

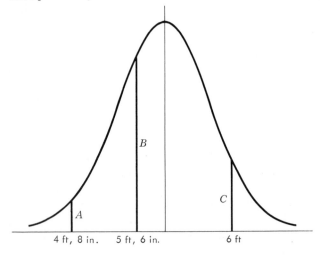

has a quite definite meaning. Any line erected from the horizontal to the curve represents the probability or chance that the variable (height, in this case) will take on the value of the point on the horizontal line from which the line rises. Put another way, the height of each of our lines *A, B,* and *C* is actually the chance that men will be as tall as the point at the bottom of the line; the longer the line, the better the odds.

For example, let us look closely at line *A.* It was erected from a point on the horizontal equal to 4 feet, 8 inches; therefore, the height of line *A* is the probability that adult males will be 4 feet, 8 inches, tall. Notice that this line *A* is quite short; this is understandable as very few people *are* 4 feet, 8 inches, tall. Now notice line *B.* This line represents the probability that adult males will be 5 feet, 6 inches, tall. It is quite a bit higher than line *A,* indicating only that there are more people 5 feet, 6 inches, tall than there are those who are 4 feet, 8 inches, tall. Line *C* represents the probability that adult males will be 6 feet tall. It is somewhat shorter than line *B,* indicating that there are fewer people who are 6 feet than there are those who are 5 feet, 10 inches. We can all confirm this through our own personal daily observations of those around us. Finally, notice the line at the average height. This is the longest line which we could have drawn under the curve, indicating that there are more people 5 feet, 9 inches, tall than there are of any other height; we usually refer to this height as the most likely value.

Notice that these distributions are symmetrical; that is, the area on the right-hand side of the average is exactly equal to the area on the left-hand side of the average. What does this mean? That the distribution of people taller than average is the same as the distribution of people shorter than average.

As we observed at the start of this section, not all distributions are symmetrical. A distribution that is not symmetrical is drawn in Figure 4–5. Incidentally, this distribution is generally called a beta distribution. We can quickly see that this distribution is not like the normal distributions we have been discussing; it is not symmetrical because the distance between point *P* and the most likely value is much larger than the distance between point *Q* and the most likely value. If the curve in Fig. 4–5 *were* a true

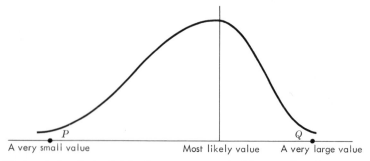

A very small value   Most likely value   A very large value

Figure 4–5. Beta distribution, unsymmetrical in one direction.

representation of the height of adult males, we would find more tall men in the world than short men.

We could, of course, draw the same beta distribution but reverse the shape of the curve to illustrate the other possibility, as in Figure 4–6. If Figure 4–6 *were* an accurate representation of the heights of adult males in the world, we would find that there are more short men than tall men. We know from personal observation, however, that neither Figure 4–5 nor Figure 4–6 is an accurate representation of the height of adult males. We would be surprised if the height of adult males were other than normally distributed, as there are just about as many short men as there are tall men. *But* that does not mean that all occurrences in this world have to be normally distributed.

Let us take an example directly from PERT. Suppose we are estimating how long it will take us to dig a basement for a house

Figure 4–6. Beta distribution, unsymmetrical in one direction.

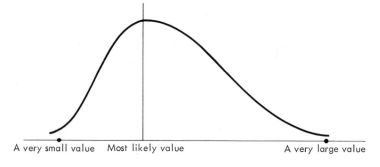

A very small value   Most likely value   A very large value

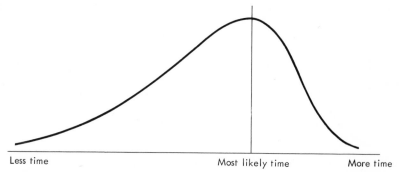

Less time   Most likely time   More time

Figure 4–7. Beta distribution for the time required to dig a basement.

by hand. We know that the weather will not be perfect, and we know that we cannot dig effectively when the ground is wet. Thus, we know for certain that if there is rain, we will no doubt fall behind our schedule for completion—the basement will require more time than we had estimated. Because we know of no development or occurrence which would counterbalance the effect of rain, our prediction about the chances of finishing this simple basement project might be pictured on a probability distribution such as that shown in Figure 4–7. The significance of this beta distribution is that it shows less chance of finishing early than of finishing late. The possible values of the variable "days required to dig" appear to be grouped nearer the "more time" side, whereas the probability lines that we might erect in the left area, the "less time" side, would all be quite short.

If we were to talk about another kind of PERT project, we could illustrate the same kind of beta distribution first illustrated in Figure 4–6. Let us assume now that we are talking about a project involving some luck; assume that we are digging a well and the chances are 7 to 1 in favor of hitting water. With these odds we would normally expect that one hole would suffice, that our first hole would be a wet hole. We know, of course, that there is a chance of digging a dry hole, and indeed a chance of five dry holes in a row, but because these chances are so small, our mental picture of our circumstances would be illustrated as in Figure 4–8. If the estimator, just to be realistic or conservative, allowed extra time for the possibility that we might hit a dry hole or two, we

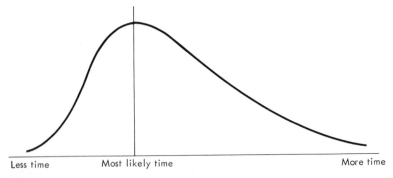

Less time   Most likely time   More time

Figure 4–8. Beta distribution for optimistic estimator.

would have excellent chances of beating his estimated time. The interpretation of Figure 4–8 is simply that we feel there is more chance of finishing in less time than predicted than there is of exceeding the time estimate.

There are, of course, many cases which we could present as further illustrations of both forms of the beta distribution, but the two illustrated will suffice at this point.

The originators of PERT were faced with the problem of finding a particular kind of distribution; they wanted a distribution:

1. With a small probability of reaching the most optimistic time (shortest time)
2. With a small probability of reaching the most pessimistic time (longest time)
3. With one and only one most likely time which would be free

Figure 4–9. Beta distribution, plain.

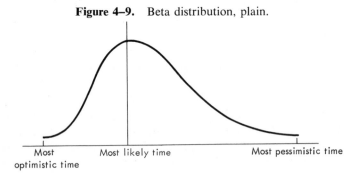

Most   Most likely time   Most pessimistic time
optimistic time

to move between the two extremes mentioned in 1 and 2 above

4. In which the amount of uncertainty in the estimating can be measured

The beta distribution was picked because it contained all four of these attributes. Let us now repeat a beta distribution (Figure 4–9) with the more usual designations under the curve and satisfy ourselves that it meets all four of the criteria just listed.

CRITERION 1

There is only a small probability of completing the project in the most optimistic time. Note that a vertical line erected at the most optimistic time (such lines are commonly called ordinates) would be short indeed, and as we have demonstrated, the length of such a line *is* the probability associated with time at that point.

CRITERION 2

The probability of completing the project in the most pessimistic time is also small. Here, too, an ordinate erected at the most pessimistic time is a short line, denoting a very small probability that the project will actually require that much time.

CRITERION 3

Clearly, there is one and only one most likely time. In this case it lies nearer to the most optimistic time, implying that the project manager or the person who has done the estimating feels that the chances are greater that the project will take less time than estimated than that it will take more time. He feels that the most likely time is nearer the most optimistic time (shortest time) than nearer the most pessimistic time.

CRITERION 4

To this point we have said nothing about measuring the uncertainty in the time estimates involved in PERT. Treatment of this particular idea will have to wait until we have become more familiar with some of the fundamental ideas in the entire PERT system. You may be assured, however, that the beta distribution meets this criterion. This fact will be demonstrated later, in Chapter 7, "Probability Concepts."

The originators of PERT, in selecting the beta distribution as appropriate for most projects on which they would use the PERT technique, were faced with the problem of computing one single time value which would be accepted as "the" time for the entire activity; they wanted *that one* of the many possible time values which would most nearly represent or reflect the expected time. Obviously, we cannot, in PERT, deal with three times for each activity; we cannot deal simultaneously with the most optimistic time, the most pessimistic time, and the most likely time. We have to select one figure which is representative of the time we think the activity will take and work with that time alone. The problem for the originators of PERT was how to combine these three time estimates to form a single time value.

Let us look briefly at the meaning of these three time estimates and then show how the three are combined to form the one time value used for the PERT network.

MOST OPTIMISTIC TIME

The most optimistic time is the particular time estimate that has a very small probability of being reached, a probability of 1 in 100. This particular time estimate represents the time in which we could finish a project if everything went along perfectly with no problems, no adverse weather, etc. We know this would be *most* unusual, but it could happen; thus, the probability of 1 in 100.

MOST PESSIMISTIC TIME

The most pessimistic time is another particular time estimate that has a very small probability of being realized, once again a probability of 1 in 100. This particular time estimate represents the time it might take us to complete a particular activity if everything went wrong, if we were plagued by adverse weather, breakdowns, bad luck, etc. We know that this, too, will not be the usual case, but it too could happen, and thus we should at least give this time *some* weight in our deliberations and computations.

MOST LIKELY TIME

The most likely time is the particular time that, in the mind of the estimator, represents the time the activity would most often require if the work were done again and again under identical conditions.

A few comments about these estimates are now in order. The most optimistic time estimate is usually represented by the letter a, the most likely time by the letter m, and the most pessimistic time by the letter b. With respect to the most pessimistic time, the estimator does not normally include in his deliberations the possibility of floods, fires, and "acts of God." Although these may indeed occur, the probability of their occurrence is less than 1 in 100.

After these three time estimates have been made, they must be combined into a single workable time value. This is done algebraically, using a weighted average derived by statisticians. Although we cannot in a book such as this go through all the complicated statistics and algebra necessary to prove the validity of the weighted average, we can make some commonsense observations on why the final formula comes out the way it does.

In the first place, in computing a weighted average we would not give the same weight to the most pessimistic time as we would give to the most likely time. There is much more of a chance that the project will be completed near the most likely time than there is that it will be completed near the most pessimistic time. Therefore the most likely time m must be weighted much heavier than the most pessimistic time b.

By the same token, the weight given to the most likely time must be much greater than the weight given to the most optimistic time; there is much more of a chance that the project will be completed near the most likely time m than there is that that it will be completed near the most optimistic time a.

Finally, there is about the same chance (1 in 100) that the project will be completed as late as the most pessimistic time as there is that it will be completed as early as the most optimistic time (1 in 100). It is no surprise, then, that these two values, a and b, are given the same weight in the algebraic formula. The formula is

$$t_e = \frac{a + 4m + b}{6}$$

where t_e = expected time for the activity, or expected elapsed time

This formula states that the expected elapsed time for an activity equals the most optimistic time plus 4 times the most likely time plus the most pessimistic time—all divided by 6. The most likely time has been given the highest weight, as stated earlier, but each of the other times has been given some weight so that we will not forget there is a small possibility that the time required to complete the project may rise to the most pessimistic time or, conversely, fall to the most optimistic time.

Now that we have been introduced to the fundamentals of how to compute t_e (the expected elapsed time), let us try our skill

Table 4–4. Examples of Calculation of Expected Elapsed Time

MOST OPTIMISTIC TIME	MOST LIKELY TIME	MOST PESSIMISTIC TIME	t_e
2	8	11	7.5
1	4	6	3.8
3	9	14	8.8
4	7	10	7.0
6	15	20	14.3
7	9	12	9.2
13	20	25	19.7
6	9	13	9.0
2	4	7	4.2
15	30	45	30.0

on a few examples. In Table 4–4 the three time values are given and t_e is calculated. All time values are in weeks; t_e will, of course, be calculated to the nearest decimal place exactly as the time estimates were.

One final note will complete our treatment of time considerations in the use of the PERT technique. Let us start with the three time estimates for a project and plot them on a beta distribution. These time estimates are as follows:

$$a = 2 \qquad m = 4 \qquad b = 13$$

Using the weighted average method

$$t_e = \frac{a + 4m + b}{6}$$

we come up with an expected elapsed time of 5.2 weeks. We can then plot the three time estimates and the resulting t_e on a beta distribution as in Figure 4–10. In this case t_e came out to be to the right of the most likely time. This is certainly understandable because the estimator felt that if things went right he could cut off 2 weeks from the most likely time $(4 - 2 = 2)$, but he also felt that if things went badly, the work would require 9 weeks more than the most likely time $(13 - 4 = 9)$. Thus, the 13-week pessimistic estimate pulled the expected elapsed time farther to the right on the distribution.

Figure 4–10. Beta distribution showing expected elapsed time lying to the right of the most likely time.

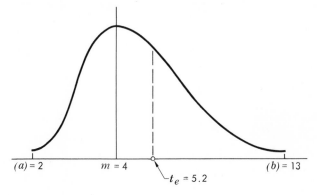

The converse of the foregoing case can just as easily be illustrated. Suppose that our three time estimates on a certain project are as follows:

$$a = 2 \qquad m = 10 \qquad b = 13$$

Using the weighted average formula

$$t_e = \frac{a + 4m + b}{6}$$

we come up with an expected elapsed time of 9.2 weeks. We then plot the three time estimates and the resulting t_e on a beta distribution as in Figure 4–11. Now we can see that t_e will lie to the left of the most likely time. This again comes as no surprise; the estimator felt that if everything went well, he could cut off 8 weeks from the most likely time ($10 - 2 = 8$), but that if everything went badly, only 3 weeks would be added to his time ($10 + 3 = 13$). This means essentially that he is a bit optimistic and thus the t_e value moves toward the most optimistic value and away from the most pessimistic value.

A final note is in order about the statistical significance of expected elapsed time. Expected elapsed time t_e is nothing more than the expected value of the distribution; that is, it represents the particular time value (expressed in weeks in this case) that has both a 50-50 chance of being exceeded and a 50-50 chance of being met. When we compute t_e, we are saying there is the

Figure 4–11. The converse of Figure 4–10.

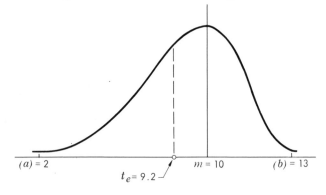

same chance the actual time taken will be *greater* than t_e as there is it will be *less* than t_e. Another way of saying the same thing is that if we erect a perpendicular line at the t_e value, half the area under the curve will lie to each side of this line. In a normal, or bell-shaped, distribution, the expected value would be the average (which is also the most likely value), but in our beta distribution, the expected value t_e will lie either to the right or to the left of the most likely value depending upon the three time estimates.

Studies have been made concerning the accuracy and validity of computation of t_e using the formula discussed in this section. One study, published a few years ago, showed that in most PERT situations, the error in t_e from calculating it with the formula we use was small enough to make the method quite satisfactory in most cases.*

* K. R. MacCrimmon and C. A. Ryavec, "An Analytical Study of the PERT Assumptions," Memo RM-3408-PR, RAND Corporation, Santa Monica, Calif., December, 1962.

5

Networking Principles

Combining networks

The fundamental concept of each **PERT** network is the activity-event relationship we have been discussing. No matter how simple or how involved the network may be, no matter how many individual projects are involved, no matter how much time is required for each phase or for the total project, we can still express the undertaking as a network consisting of only events and activities. We shall begin with two very simple networks and then combine these to form a slightly more involved network. This is the procedure followed in building up smaller subnetworks to form the total network.

Suppose that our project involves the design and production of a prototype airplane, the first of its kind. Further, assume that it will be a new

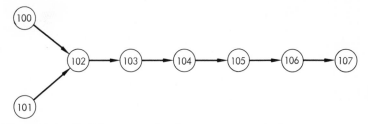

Figure 5–1.   Building an engine for a prototype airplane.

light plane involving rather conventional engine and airframe construction. Both sections of our company (engine and airframe) have prepared their PERT networks. The network for the engine section is shown in Figure 5–1. A detailed description of the activities and events in this network is given in Table 5–1.

Of course, the airframe section work must go on in order that a completed airframe will be ready for engine installation.

Table 5–1. Description of Activities and Events of the Network in Figure 5–1

ACTIVITY NO.	ACTIVITY DESCRIPTION	EVENT DESCRIPTION
100-102	Prepare plans and specifications for engine	102—Bids and specifications completed; sources located
101-102	Locate sources for engine procurement	102—Bids and specifications completed; sources located
102-103	Forward plans and specifications to eligible bidders	103—Plans and specifications received by bidders
103-104	Analyze bids received and award contract	104—Contract signed
104-105	Maintain procurement follow-up action with successful bidder	105—Completed engine received
105-106	Test engine	106—Test completed
106-107	Install engine in airframe	107—Engine mated to airframe

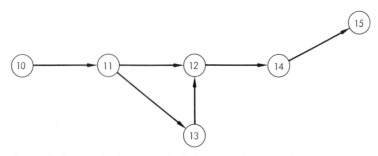

Figure 5–2. Designing and building an airframe for a prototype airplane.

We shall assume here that the company has elected to design and manufacture its own airframe and that the process is described by the network in Figure 5–2. Table 5–2 gives a detailed description of the activities and events in the network shown in Figure 5–2.

When we look at the networks in Figures 5–1 and 5–2 and note the descriptions of the activities and events in Tables 5–1

Table 5–2. Description of Activities and Events of the Network in Figure 5–2

ACTIVITY NO.	ACTIVITY DESCRIPTION	EVENT DESCRIPTION
10-11	Perform initial design work	11—Airframe design completed
11-12	Procure required materials for airframe	12—Materials received
11-13	Prepare necessary jigs and fixtures for assembly of airframe	13—Jigs and fixtures completed
13-12	A "dummy" activity showing that jigs and fixtures must be available before assembly can begin	
12-14	Assemble airframe	14—Airframe completed
14-15	Install engine in airframe	15—Engine mated to airframe

and 5–2, one fact is obvious: Events 15 and 107 are the same event—mate airframe and engine. When an event such as event 15 or 107 is common to two or more networks, it is called an *interface event* and is noted with a special designation such as

⬡ instead of merely ◯ ; interface events may also be represented by ⬤

Interface events are common when smaller networks are combined to make up the master network (the network for the complete project). For large, complex networks, thousands of activities are required to reflect completely and accurately what is involved. Having as many as 100 smaller networks combined in one larger network is not at all unusual.

To join the two networks presented in Figures 5–1 and 5–2, we need only superimpose the interface events (15 and 107) on each other. This is accomplished in Figure 5–3. The interface event has been given a special designation and size. Notice that the network shown in this figure has no single network-beginning event. If it looks a bit disconnected, we could define a network-

Figure 5–3. Master network formed by combining Figures 5–1 and 5–2.

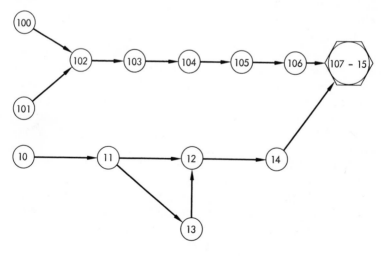

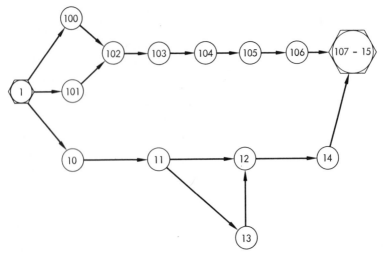

Figure 5–4. Figure 5–3 with a network-beginning event added.

beginning event (number 1), and put it out in front of the network. We would thus have created three new activities, 1-10, 1-100, and 1-101, all of which could be described as "project go-ahead notice given to airframe or engine sections." If we did this, the resulting network would appear as in Figure 5–4. Event 1 has been shown as an interface event since, technically speaking, it is now common to both networks.

In joining networks through the use of interface events, we have used a rather arbitrary event-numbering system simply to separate the two networks in the reader's mind. There are many numbering conventions in use in networking today. Some of these are required by computer needs; others simply add clarity to the process. Any convention that adds to the usefulness of this planning technique should be welcomed and used.

Not all networks are combined as easily as those shown in Figures 5–1 and 5–2; however, the principle of the interface event is still used in the combining process. In Figure 5–5, three individual networks are shown first; the three are then combined by recognizing and using the interface events. It is not necessary to describe the activities in this example.

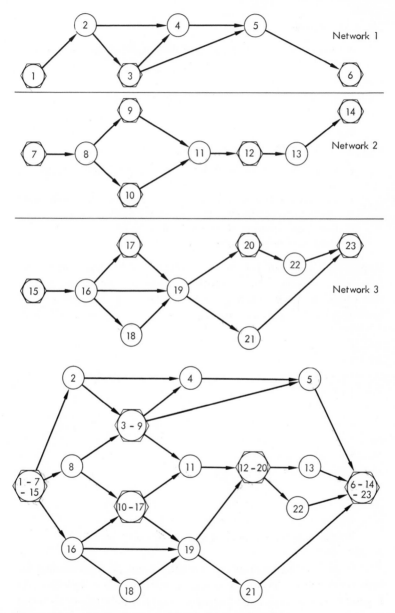

Figure 5–5. Method of combining three networks.

Zero-time activities

A second concept that we shall find useful in the preparation of our PERT networks is the zero-time activity. Although it may seem unrealistic that certain activities take no time whatsoever, a few simple illustrations should indicate that this *can* happen and indeed *must* happen in certain cases.

Suppose that we are building a house and that the first five steps we can isolate are the following:

1. Clear the site (remove trees, rocks, etc., and prepare for building)
2. Lay out the project (put up stakes, measure distances, etc.)
3. Construct outside masonry walls up to floor level
4. Construct interior piers up to floor level
5. Construct floor of house

Steps 4 and 5 in our list could be illustrated as shown in Figure 5–6. The function of the interior piers is to support the lumber that forms the floor, when the distance between the outside walls is more than the lumber can effectively span without bending.

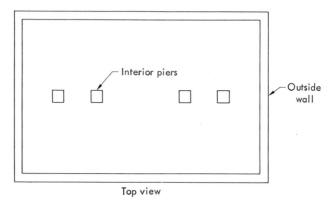

Top view

Figure 5–6.   Interior piers and outside walls of a house.

If we wanted to draw up a PERT network representing the first six activities in this project, it might look something like Figure 5–7. We have added expected elapsed time values in terms

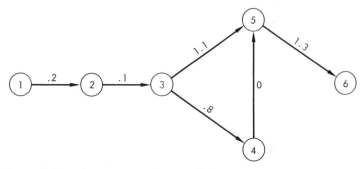

Figure 5–7.   First six activities in building a house.

of weeks for each activity. A formal PERT listing of our activities and definitions is given in Table 5–3.

The activities on the Figure 5–7 diagram are self-explanatory with the exception of activity 4–5 which requires no time. We might ask "If it takes no time, why include it in the PERT network?" There is good reason for its inclusion. Notice that activities 3–5 and 3–4 can go on simultaneously; the masonry contractor can construct the outside walls *and* interior piers at the same time if he so desires. One does not depend upon the other in any way. But now notice that activity 5-6 (construct floor) *cannot* begin until both the exterior walls and the interior piers have been completed. If we tried to construct a floor without completed interior piers, the floor would bend and fail; if we tried to lay the floor without completed exterior walls, there would be nothing to support the outside at all. Thus, both activity 3-5

Table 5–3. Definitions of First Five Activities in Building a House

ACTIVITY	DEFINITION
1-2	Clear site
2-3	Lay out house
3-5	Construct outside walls up to floor level
3-4	Construct interior piers up to floor level
4-5	(A zero-time activity)
5-6	Construct floor

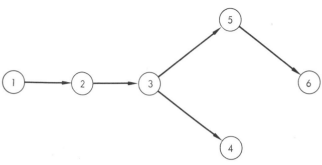

Figure 5–8.   Figure 5–7 with the zero-time activity omitted.

and activity 3-4 must be completed before activity 5-6 can begin. This is exactly what the zero-time activity 4-5 tells us; it says simply that event 4 (completion of interior piers) and event 5 (completion of outside walls) must both take place before activity 5-6 (construct floor) can begin.

Suppose we were to redraw the network without the zero-time activity, as in Figure 5–8. Now what does this network tell us? Just that activity 5-6 (construct floor) can proceed as soon as activity 3-5 (construct outside walls) ends. But what about the interior piers? If they support the floor, it is obvious we cannot go ahead with the floor until we finish them. Placing a zero-time activity between events 4 and 5 simply indicates an order of precedence and illustrates dependence of one event upon another. This is the function of the zero-time activity. A zero-time activity can be defined formally as one that constrains an event from happening until another event has occurred. Some sources represent zero-time activities with a dotted line, but we shall use the more normal convention of a solid line with a time value of zero.

Earliest expected date

Let us now turn to *earliest expected date,* or T_E, a PERT concept that is concerned with the time required to complete certain work. Picture a small network such as that represented by Figure 5–9. Our project manager must know when he can expect completion of the project represented by this network. Looking

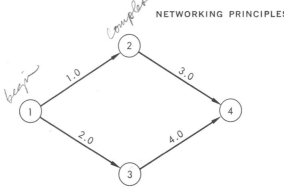

Figure 5–9. Simple network illustrating significance of T_E.

at Figure 5–9, we see that there are two paths through this net-work, path 1-2-4 and path 1-3-4. We can further see that per-formance of the work on path 1-2-4 will take a total of 4.0 work-ing weeks while performance of the work on path 1-3-4 will require a total of 6.0 working weeks. The earliest expected date for network completion is 6.0 weeks. Even if activities 1-2 and 2-4 were completed in a total of 4.0 working weeks, we would still have to wait until 6.0 weeks had passed for event 4 to be completed, as the lower path through the network (path 1-3-4) will require a total of 6.0 working weeks for completion.

We can say, then, from Figure 5–9 that the earliest expected date for completion of event 4 is 6.0 weeks. When we say the earliest expected date is 6.0 weeks, we mean that completion can be expected no earlier than 6.0 weeks from today *if* we begin today. If the starting date were January 1, completion could be expected no earlier than February 12. The term "earliest expected date," thus, is associated with *events* in PERT and refers to the earliest time at which we may expect them to occur.

The earliest expected date of an event is computed by cal-culating the *longest* path from the network-beginning event to the event in question; this can be the final event or some other event. Even if one of the paths can be completed in a shorter time, it is ultimately the *longest* path which will determine the amount of time necessary.

Now that we are clear on definitions and have worked through a very simple example, let us try a more difficult problem.

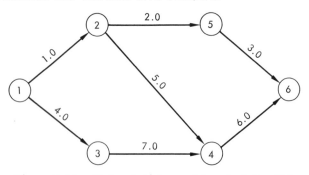

Figure 5–10. Network to be used in calculating T_E's.

Remember the definition and use of earliest expected date; if the calendar date of the network-beginning event is given, the earliest expected date of any event in the network can be figured as a calendar date. If no calendar date for the network-beginning event is given, the earliest expected date of any event in the network must be figured in terms of elapsed weeks.

In Figure 5–10 we have illustrated a slightly more complicated PERT network. Suppose that our project manager wants to know just how soon he can expect completion of the work represented by this network. To calculate the earliest expected date of event 6 (the network-ending event) let us start summing the times for all the paths through the network. And just how many paths are there?

1-2-5-6	1.0 + 2.0 + 3.0 =	6.0 weeks
1-2-4-6	1.0 + 5.0 + 6.0 =	12.0 weeks
1-3-4-6	4.0 + 7.0 + 6.0 =	17.0 weeks

We see that the earliest expected date of event 6 is 17.0 weeks from the time we begin the network. True, two other paths through the network can be completed in less time than 17.0 weeks, but even if they were, completion of the entire network would have to wait until the three activities on the longest path had been completed: activity 1-3 requires 4.0 weeks, activity 3-4 requires 7.0 weeks, and activity 4-6 requires 6.0 weeks. Thus, the earliest we may expect completion of this network is 17.0 weeks after starting.

Now suppose that our manager wants to know how soon he

can expect completion of an event which is not the ending event of the network. For example, when can we expect event 4 to be completed? Use the same rule as before: Sum the paths from the network-beginning event to the event in question, and take the longest path. From the network-beginning event to event 4 there are only two paths, 1-2-4 and 1-3-4. The first requires 6.0 weeks for completion and the second requires 11.0 weeks. We must therefore reason that the earliest expected date for event 4 is 11.0 weeks. Remember that although path 1-2-4 could be completed in 6.0 weeks, we would still have to wait until the two activities on the longer path (activities 1-3 and 3-4) had been completed, and 11.0 weeks are necessary to complete them.

The earliest expected date T_E for each event in our network is computed by using the same rule, by taking the longest path from the network-beginning event to each of the other events, as in Figure 5–11. Although the T_E for the network-beginning event is meaningless, the convention is to list it as zero, which we have done. The significance of T_E is simply that we are able to compute the date on which certain work will be completed. If this date is unsatisfactory, we may have to make some adjustment in our work schedule or bring in more men and/or equipment, but at this initial point, we are at least able to predict with some certainty when events will occur in the future.

Figure 5–11.   Figure 5–10 with the earliest expected date added for each event.

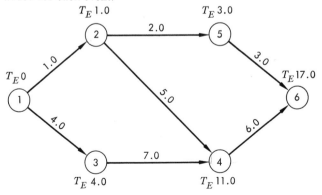

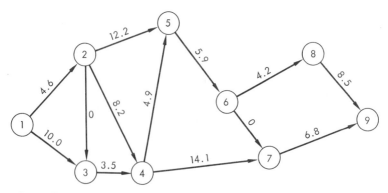

Figure 5–12. Complex network to be used in calculating T_E's.

Let us now try a fairly large network and see how we get along. In Figure 5–12 we have a network which contains 9 events; you will notice also that it contains two zero-time activities. Let us start with the network-beginning event and calculate the earliest expected dates for each of the events on the network. Using the normal convention which assigns a T_E of zero to event 1, we can proceed directly to event 2. As there is only one path from the network-beginning event to event 2, the earliest expected date must be 4.6 weeks. Event 3, however, is somewhat different. There are two paths from the network-beginning event to event 3, paths 1-2-3 and 1-3; the longer of these paths is path 1-3 which requires 10.0 weeks for completion. Hence, the T_E of event 3 is 10.0 weeks. Now, to calculate the T_E of event 4, we do not have to go all the way back to the network-beginning event unless we want to. There are only two ways to get to event 4; one is from event 2, the other is from event 3. Knowing the T_E's of events 2 and 3, we need simply to add to these the expected elapsed times of activities 2-4 and 3-4 to get the possible answers for the T_E of event 4. The earliest expected date of event 3 is 10.0 weeks, and activity 3-4 requires 3.5 weeks; one answer for the T_E of event 4 is therefore 13.5 weeks. There is still another possibility, however. The T_E for event 2 was 4.6 weeks, and as activity 2-4 consumes 8.2 weeks, we add these together and get another unit of time, 12.8 weeks. Because T_E is the *longest* path from the network-beginning event to the event in question, we are forced to accept 13.5 weeks as the correct T_E for event 4.

The procedure from here on is repetitive. To find the T_E for event 5, take the earliest expected dates for events 2 and 4 (the only possible ways to get to event 5) and add to them the expected elapsed time of the paths from events 2 and 4 to event 5. The T_E for event 4 is 13.5 weeks; to this we add 4.9 weeks to get one answer, 18.4 weeks. The other approach involves adding 12.2 weeks to the T_E for event 2, which was 4.6 weeks, for a total of 16.8 weeks. The longer path is certainly 18.4 weeks, the correct T_E for event 5.

Because there is only one path from event 5 to event 6, calculation of the T_E for event 6 is merely a matter of adding 5.9 weeks to the T_E for event 5 (18.4 weeks) to get the answer of 24.3 weeks. Now go to event 7. There are two possible ways to reach event 7, one from event 6, the other from event 4. Adding the zero-time activity to the T_E of event 6 still yields 24.3 weeks; adding 14.1 weeks to the T_E for event 4 yields 27.6 weeks. The true answer is, of course, 27.6 weeks.

The T_E for event 8 is another simple one; there is only one path—through event 6. Adding the T_E for event 6 (24.3 weeks) to the expected elapsed time for activity 6-8 (4.2 weeks) yields

Figure 5–13. Figure 5–12 with the earliest expected date added for each event.

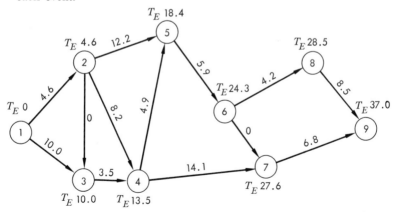

28.5 weeks as the T_E for event 8. From events 7 and 8 to the network-ending event there are two paths. Adding 8.5 weeks to the T_E for event 8 yields 37.0 weeks as one alternative. Going the other way we add 6.8 weeks to the T_E for event 7 (27.6 weeks) and get 34.4 weeks. We must choose 37.0 weeks as the correct answer for the earliest expected date for the network-ending event. The network is redrawn in Figure 5–13, and the correct T_E values for each event are included.

Ponder for a moment the significance of these earliest expected date figures. What do they really mean? The project manager now has a schedule of when he can expect completion of each of the activities in the network. This schedule in no way guarantees completion time, but it does enable the manager to do a week-by-week comparison of actual performance against scheduled performance. He can thus ascertain at any time whether he is ahead of, behind, or exactly on schedule. Suppose, for example, that we have been working on the project illustrated in Figure 5–13 for exactly 9.0 weeks. We get a report from the field that activity 1-3 will not be completed for 3.0 more weeks. We are at least in a position to know that we are now about 2.0 weeks behind schedule with activity 1-3 and that some action probably must be taken.

Effects of zero-time activities

Let us return for a moment to the concept of zero-time activities. Their significance can best be understood by working through the following two examples. In one, the addition of a zero-time

Figure 5–14. Network used to illustrate effect of zero-time activity.

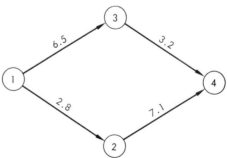

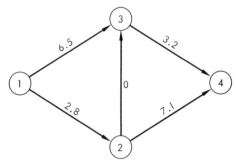

Figure 5–15. Figure 5–14 with a zero-time activity added. The zero-time activity in this case has no effect on the earliest expected date.

activity has no effect upon the earliest expected date of the network-ending event; in the other, the addition of the zero-time activity extends considerably the earliest expected date of the network-ending event.

Consider the network shown in Figure 5–14. Using our method for determining T_E, we can see that the earliest expected date for completion of this network is 9.9 weeks, the sum of the activity times on the longer path through the network. Now let us add a zero-time activity from event 2 to event 3, indicating only that work on activity 3-4 cannot go on until event 2 has been completed. This is done in Figure 5–15.

What is the significance of this inclusion? Simply that there are now three paths through the network—paths 1-3-4, 1-2-3-4, and 1-2-4. The total times required to complete the work on these paths are 9.7 weeks, 6.0 weeks, and 9.9 weeks, respectively. Thus, the earliest expected date of the network-ending event is still 9.9 weeks. The addition of the zero-time activity (a sequence or precedence notation) has in no way affected the T_E of the network-ending event. We still expect to complete this network in 9.9 weeks.

Now let us take another network similar to the one illustrated in Figure 5–14 and try the same move. In Figure 5–16 we have another network with four events. The longer of the two paths through this network, path 1-3-4, yields a T_E value for the network-ending event of 13.5 weeks.

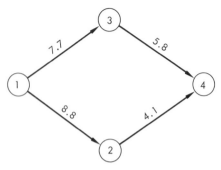

Figure 5–16. Network used to illustrate effect of zero-time activity.

Now let us add a zero-time activity as shown in Figure 5–17. In this figure there are three paths through the network—paths 1-3-4, 1-2-3-4, and 1-2-4. The new path created, 1-2-3-4, turns out to be longer than the longest previous path and thus establishes a new T_E for the network-ending event, 14.6 weeks. Whereas before we could be reasonably certain of finishing this work in 13.5 weeks, the inclusion of a precedence or sequence restraint has now changed the T_E to 14.6 weeks.

What is the practical significance of what we have illustrated? Just this: In some projects, requiring a certain sequence of work will not affect the time required to complete the entire project, whereas in others the requirement that a certain event be completed before another event may begin can have definite time ramifications. We have already seen that the time required for each individual phase of the network influences the earliest expected

Figure 5–17. Figure 5–16 with a zero-time activity added. The zero-time activity in this case affects the earliest expected date.

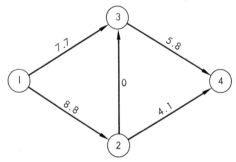

date. Now we recognize that the *sequence* of jobs which must be followed can be another determinant of that date.

Critical path

We have illustrated how the longest path through the network determines the earliest expected date of the network-ending event. This longest path is most commonly referred to as the *critical path*. ✓ The critical path is the most time-consuming path of activities from the beginning to the end of the network. Strange as it may seem at first, there can be more than one critical path through a network. Take for instance the network illustrated in Figure 5–18. There are only two paths through this network—paths 1-2-3-5-6 and 1-2-4-5-6. Summing up the expected elapsed times for the activities along these two paths yields the same answer, 12.0 weeks. Each of these paths is the longest path through the network, and both are therefore critical paths.

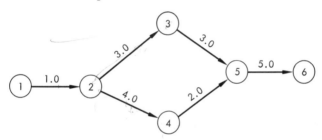

Figure 5–18.   Network with two critical paths.

We can also illustrate a network containing several critical paths *and* several noncritical paths, as in Figure 5–19. A little bit of arithmetic will illustrate that in the network in this figure there are three critical paths:

1-2-5-6-8	10.0 weeks
1-3-4-7-8	10.0 weeks
1-2-4-7-8	10.0 weeks

and other paths which are not critical:

1-3-5-6-8	9.0 weeks
1-3-5-7-8	7.0 weeks
1-2-5-7-8	8.0 weeks

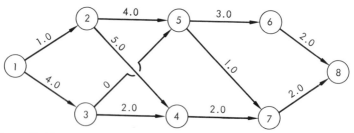

Figure 5–19. Network with critical and noncritical paths.

Notice that this network contains only eight events but has three critical paths. It is no wonder that a network of only a few dozen events complicates the problem of hand calculation and suggests the use of a computer for its solution.

Latest allowable date

The concept of *latest allowable date* has to do with the latest calendar date on which an event can take place and still not interfere with the scheduled completion date of the network. It too can be expressed as a calendar date if the calendar date of the network-beginning event is given; if not, it is expressed as the greatest number of weeks which can elapse between the network-beginning event and any event in question without delaying the scheduled completion of the network. We can illustrate this concept by referring to Figure 5–20. In this network the T_E of the network-ending event is 16.0 weeks and path 1-3-5-6 is the longest of the three possible paths through the network. Using the methods

Figure 5–20. Network used to illustrate calculation of T_E and T_L.

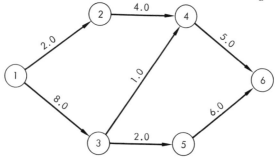

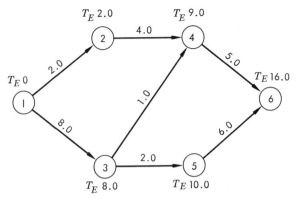

T_E 2.0 T_E 9.0

T_E 0

T_E 16.0

T_E 8.0 T_E 10.0

Figure 5–21. Figure 5–20 with the earliest expected dates added.

outlined earlier in this chapter, we have calculated the earliest expected dates for each of the six events in the network and illustrated them in Figure 5–21. Now look at event 4. We expect it to be completed 9.0 weeks after the network is begun. If we began activity 4-6 immediately thereafter, we would be ready for event 6 to occur in 14.0 weeks after the network was begun, but we see that event 6 cannot occur until 16.0 weeks after the beginning date because of other activity paths. Thus, we do not have to complete event 4 in 9.0 weeks; we could actually complete it 11.0 weeks after the network was begun and still not interfere with our scheduled network time of 16.0 weeks. This is the significance of latest allowable date. The latest allowable date (symbolized T_L) for event 4 is 11.0 weeks after the network begins because at that time we still have 5.0 weeks' work to do to complete activity 4-6 and exactly 5.0 weeks in which to do it.

Suppose that event 4 were not completed until 12.0 weeks after the network had been started. With 5.0 weeks' work still to be done (activity 4-6) we could not keep the scheduled completion date of 16.0 weeks but would run beyond by 1.0 week (12.0 weeks + 5.0 weeks = 17.0 weeks).

Now let us calculate the T_L for event 5. In order for the network to be completed on time (in 16.0 weeks) we must complete event 5 within 10.0 weeks after the network begins or we shall be late. Thus, the T_L for event 5 is the same as its T_E, 10.0 weeks.

Now let us compute the T_L for event 2. Between event 2 and

the end of the network, there is a maximum of 9.0 weeks' work to
be done, the sum of activities 2-4 and 4-6. Thus, event 2 can be
completed as late as 7.0 weeks after the network begins and still
not interfere with our scheduled completion date of 16.0 weeks.

What about event 3? Between event 3 and the network-ending
event there are two paths. One of these paths involves 6.0 weeks'
work (3-4-6); the other involves 8.0 weeks' work (3-5-6). If we
.are not to exceed our scheduled completion date of 16.0 weeks,
event 3 must be completed 8.0 weeks after the network is begun to
allow 8.0 more weeks for the accomplishment of the work on activities 3-5 and 5-6. This particular illustration indicates that although there may be more than one path between an event and
the network-ending event, the longest path determines the T_L in
every case.

The latest allowable date for event 1 (the network-beginning
event) is of course zero in this case. This simply means that if
we begin after our scheduled beginning time we will be late. We
could prove this by finding the longest path between event 1 and
the network-ending event (1-3-5-6). The sum of the expected
elapsed times along this path is 16.0 weeks, which, when subtracted
from the T_E of the network-ending event, gives zero as an answer.

The network from Figure 5–20 is reproduced in Figure 5–22
with all the T_E's and T_L's illustrated. Observe that the critical path

Figure 5–22.   Figure 5–21 with the latest allowable dates added.

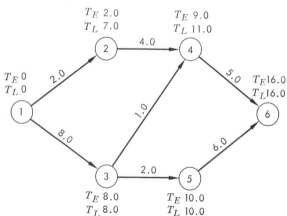

is 1-3-5-6, the longest path through the network. Observe also that for each of the events on this critical path, its T_E is exactly equal to its T_L. What does this mean? Simply that the latest allowable date on which each can be completed is exactly equal to the earliest date on which we may expect each to be completed. In simple terms, there is no time to spare; the events must be completed exactly when we have scheduled them or we will never make our scheduled time of 16.0 weeks for the project.

Now look for a minute at an event not on the critical path, event 4, for example. Although its earliest expected date is 9.0 weeks after the network begins, it does not have to be completed until 11.0 weeks after the network begins. This means we can fall behind on activity 2-4 as much as 2.0 weeks and still not jeopardize finishing the network in 16.0 weeks. Let us look at another event not on the critical path, event 2. We expect it to be done just 2.0 weeks after we begin the network, but we see it does not have to be done until 7.0 weeks after the network is started. We could really start it 5.0 weeks late and still complete the network on time, or we could fall behind 5.0 weeks while working on activity 1-2 and not interfere with the scheduled completion date. If, however, we fell behind so much that event 2 was not completed until 8.0 weeks after the network had begun, the remaining 9.0 weeks of work (activities 2-4 and 4-6) are such that the network could not be completed until 17.0 weeks after beginning, and we would thus be late.

This situation indicates another valuable use of PERT: enabling us to see where we can or must save time and where we can let the schedule slide for a while if that is advantageous. From the example we have just worked out, it is obvious that if we do not stick to our original schedule on events 3, 5, and 6 we shall never finish in 16.0 weeks, *but* we can fall behind in varying degrees on events 2 and 4 without affecting our scheduled completion date.

Before we move on to another topic, let us work with a somewhat more complicated network, illustrated in Figure 5–23, in attempting to calculate the T_E's and T_L's for each of the events. We shall begin with the earliest expected dates. As there is only one network path to event 2, its T_E is 4.2 weeks; to calculate the

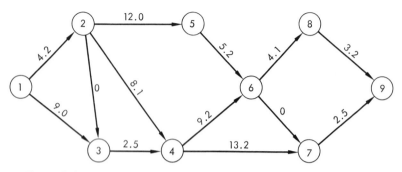

Figure 5–23. Network used to illustrate calculation of T_E and T_L.

T_E for event 3 we have two choices: path 1-2-3 (4.2 weeks) or path 1-3 (9.0 weeks); we of course choose the latter. Event 4 is reached from either event 2 or event 3, the longer path being from event 2 and yielding a T_E for event 4 of 12.3 weeks. Event 5 is reached from one direction only, so that its T_E becomes the T_E of event 2 plus 12.0 weeks, or 16.2 weeks. To get the T_E of event 6, we can add 5.2 to the T_E for event 5—or 9.2 to the T_E for event 4; the latter course yields the larger answer, 21.5 weeks. The T_E for event 7 must be either the T_E for event 6 (activity 6-7 is a zero-time activity) or the T_E for event 4 plus 13.2 weeks; the latter course yields the correct answer of 25.5 weeks. Event 8 is reached only through event 6 and its T_E must then be the T_E for event 6 plus 4.1 weeks, or 25.6 weeks. Finally, the earliest expected date for the network-ending event (event 9) must be either the T_E for event 8 (25.6 weeks) plus 3.2, or the T_E for event 7 (25.5 weeks) plus 2.5 weeks; the larger answer is 28.8 weeks. That completes calculation of all the earliest expected dates for the network events.

In Figure 5–24 the same network is shown with the earliest expected date indicated for each event. Now let us calculate the latest allowable date for each of the events in our network. Beginning with event 8, and noticing that there is only one path from event 9 back to event 8, we calculate the T_L for event 8 at 25.6 weeks. There is also only one path from event 9 back to event 7; thus, the T_L for event 7 is 28.8 − 2.5, or 26.3 weeks. There are two

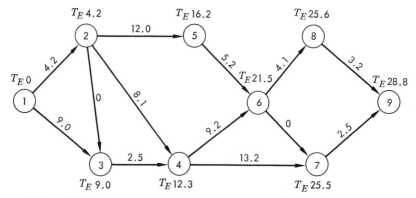

Figure 5–24. Figure 5–23 with the earliest expected dates added.

paths between event 6 and the network-ending event, paths 6-8-9 and 6-7-9. Deducting the time of the longer path from 28.8 weeks, we get 21.5 weeks as the T_L for event 6. The T_L for event 5 is obtained by subtracting the expected elapsed time for activity 5-6 from the T_L of event 6 (since there is only one path between events 5 and 6) with a result of 16.3 weeks. To obtain the T_L for event 4 we may subtract either 9.2 from the T_L for event 6 (21.5 weeks), or 13.2 from the T_L for event 7 (26.3 weeks). The smaller answer is 12.3 weeks, the T_L for event 4. As for event 3, there is only one path between it and event 4; the T_L for event 3 must therefore be $12.3 - 2.5$, or 9.8 weeks. Finally, the T_L for event 2 is obtained by working backwards from either event 5, event 4, or event 3. The smallest answer is 4.2, the correct T_L for event 2.

The entire network, including designation of all T_E's and T_L's, is illustrated in Figure 5–25. Notice that recognition of the critical path is quite easy, since the critical path will contain each of those events whose T_E is equal to the T_L; in this case the critical path is 1-2-4-6-8-9, indicated by a heavy line.

Whereas calculation of T_E's is a process of *addition* (finding the longest path from the network-beginning event to the event in question), calculation of T_L's is a process of *subtraction* (finding the longest path from the network-ending event back to the event in question). In the calculation of T_E's, when two or more network

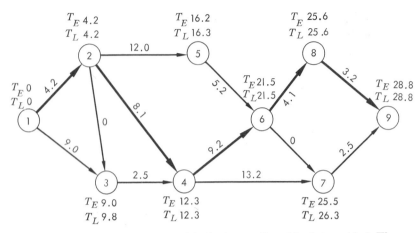

Figure 5–25. Figure 5–24 with the latest allowable dates added. The critical path is indicated by a heavy line.

paths yield different values, we always choose the larger; in the calculation of T_L's, when two or more network paths yield different values, we always choose the smaller.

Finally, here are two shortcuts which you probably have been using all the time intuitively. First let us construct a brief network, as in Figure 5–26. Now calculate the T_E for event 4; this involves choosing the longer of two paths, 1-2-4 and 1-3-4; path 1-3-4 turns out to be the longer of the two, and the T_E of event 4 is 4.0 weeks.

Figure 5–26. Network in which one event is the junction of all preceding activities, illustrating a shortcut for calculating the earliest expected date.

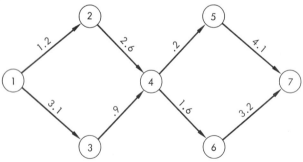

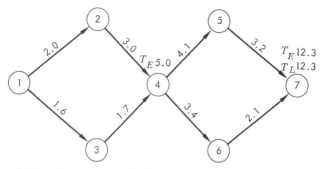

Figure 5–27. Network in which one event is the junction of all preceding activities, illustrating a shortcut for calculating the latest allowable date.

Since event 4 is the termination event of all activities that precede it, we need *not* go all the way back to the network-beginning event to calculate the T_E's of events beyond this point. For instance, the T_E for event 5 must be 4.0 (the T_E for event 4) plus .2 week, the expected elapsed time for activity 4-5. By the same reasoning, the T_E for event 6 must be 4.0 (again the T_E for event 4) plus 1.6 weeks, the expected elapsed time for activity 4-6. This is a useful shortcut whenever you are confronted by an event which is the terminus of all preceding activities.

A second shortcut involves the same idea in reverse. Look at the network in Figure 5–27. When calculating the T_L's of events to the left of event 4, you need not go all the way back to the end of the network once the T_L of event 4 has been established as 5.0. The T_L's of events 2 and 3 can be calculated by using this figure of 5.0 as a base point; for instance, the T_L for event 2 must be $5.0 - 3.0$, or 2.0, for if event 2 is completed later than 2.0 weeks after the network begins, there will not be sufficient time left for the completion of the three activities which follow event 2, namely, activities 2-4, 4-5, and 5-7, with a combined time of 10.3 weeks required for completion.

Slack

The final topic for consideration in this section will be *slack time*. The idea here is exactly what the term implies, "time to

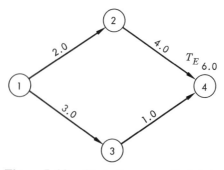

Figure 5–28. Slack time on path 1-3-4.

spare." A look at the network in Figure 5–28 will illustrate this point.

Path 1-2-4 is the longer path through the network and requires 6.0 weeks for its completion; hence, the T_E of the network-ending event must be 6.0 weeks. Path 1-2-4 is then the critical path. But what about the events on path 1-3-4? Suppose we are a week late in finishing up the work required for activity 1-3; does it really matter? Even if activity 1-3 is not completed until 4.0 weeks after the network begins, we are still in good shape because with only 1.0 week's work remaining (activity 3-4), we can finish the lower path in a total of 5.0 weeks. At that time we still would have to wait 1.0 week for the work on the other path to be completed. Actually, event 3 could be 2.0 weeks late and *still not extend* the T_E of the network-ending event.

We say, then, that there is slack on path 1-3-4. This means that we need not be worried if we fall slightly behind on this path. On the other hand, if we fall behind on the critical path, 1-2-4, the T_E of the network-ending event will fall behind to the same extent. Why? Because there is *no* slack on the critical path. From a practical point of view, slack means more time to work, less to worry about, and a chance to transfer men, machinery, or supervision to an activity that lies on the critical path. Knowing which events have slack time is of paramount importance to management.

Slack time is defined by the equation:

$$S = T_L - T_E$$

Although this may appear a bit formal, the logic behind it is quite clear. It says simply that if you take the latest allowable date on which you can finish and from this subtract the earliest expected date on which you may finish, the time that remains is slack or surplus. For example, if an intermediate event is expected to be finished by the 10th week (its $T_E = 10.0$ weeks) and if that same event could be finished by the 12th week without interfering with the T_E of the network-ending event (its $T_L = 12.0$ weeks), then that event has 2.0 weeks to spare; there is 2.0 weeks' slack:

$$S = T_L - T_E$$

or

Slack $= 12.0 - 10.0 = 2.0$ weeks

Now let us look at a slightly more complicated network, such as that in Figure 5–29, and calculate the slack time for all the events on the network. By using the methods outlined earlier, we have already calculated the T_E's and T_L's for all events in the network. By using the formula $S = T_L - T_E$, we can calculate the slack time for each event as shown in Table 5–4. Notice that only three events have any slack time: events 3, 4, and 6. They have a combined total of 6.0 weeks of slack time. From a practical point

Figure 5–29.   Network with slack time on three noncritical paths.

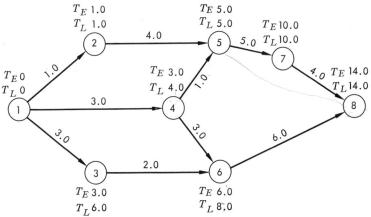

Table 5–4. Calculation of Slack Time for the Events in Figure 5–29

EVENT	T_L	—	T_E	=	SLACK
1	0	—	0	=	0
2	1.0	—	1.0	=	0
3	6.0	—	3.0	=	3.0
4	4.0	—	3.0	=	1.0
5	5.0	—	5.0	=	0
6	8.0	—	6.0	=	2.0
7	10.0	—	10.0	=	0
8	14.0	—	14.0	=	0

of view this means that we could fall 6.0 weeks behind somewhere on these three events and not interfere with completion of the project on time, at its earliest expected date. From a still *more* practical point of view, it obviously means also that we could switch resources (men, machinery, materials) to the critical path and perhaps shorten the total project time. This is the central point in **PERT**, its real reason for being and use.

Notice also from Figure 5–29 that the events on the critical path, events 1, 2, 5, 7, and 8, have no slack time; the T_L for each

Figure 5–30. Network with slack time on the critical path.

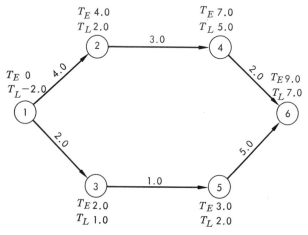

event exactly equals the T_E for that event, and therefore slack is zero. Can there be slack on the critical path? Yes. Take for instance the network in Figure 5–30. In this network we have arbitrarily stated that the T_L of the network ending event is 7.0 weeks, 2.0 weeks less than the T_E for that event. This says simply that we have allowed the project 2.0 weeks less time than it is expected to take. Though this may at first blush seem ridiculous, think of all the situations where a project has been completed (through adequate planning, good control, and maybe a little luck) in less time than originally scheduled for its completion. At any rate, when we use the formula $S = T_L - T_E$ for the calculation of slack time, we get the answers given in Table 5–5.

A glance back at the network itself will reassure you that the critical path is indeed 1-2-4-6, but you cannot describe the critical path in this instance as having no slack. The critical path in our network *has* slack; it is *negative* slack, to be sure (negative slack simply means we are behind), but slack just the same. Look for a moment at the events not on the critical path; they too have slack, but not so much negative slack. This means they are not so far behind: they are only 1.0 week behind, whereas the events on the critical path are 2.0 weeks behind. A better definition then of the critical path with respect to its slack time is: the path with the least algebraic value of slack. Thus, in our example, being 2.0 weeks behind is worse than being 1.0 week behind, or to use the proper algebraic signs, a path with −2.0 weeks slack is certainly more critical than a path with −1.0 week of slack.

Table 5–5. Calculation of Slack Time for the Events in Figure 5–30

EVENT	T_L	−	T_E	=	SLACK
1	−2.0	−	0	=	−2.0
2	2.0	−	4.0	=	−2.0
3	1.0	−	2.0	=	−1.0
4	5.0	−	7.0	=	−2.0
5	2.0	−	3.0	=	−1.0
6	7.0	−	9.0	=	−2.0

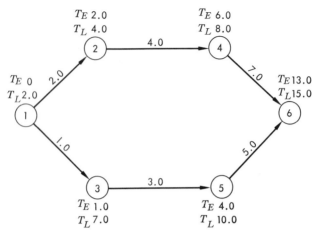

Figure 5–31. Network illustrating T_L greater than T_E.

One additional example will help clarify this very important point. Look now at the network illustrated in Figure 5–31. We have arbitrarily allowed too much time in this particular case; we set the T_L of event 6 at 15.0 weeks when the corresponding T_E is only 13.0 weeks. When we use the formula $S = T_L - T_E$ for the calculation of slack, we get the results shown in Table 5–6. Let us test our claim that the critical path is the path with the least algebraic value of slack. The slack on the critical path is 2.0 weeks while that on the other path (1-3-5-6) is 6.0 weeks. Obviously, the critical path has less slack.

Table 5–6. Calculation of Slack Time for the Events in Figure 5–31

EVENT	T_L	—	T_E	=	SLACK
1	2.0	—	0	=	2.0
2	4.0	—	2.0	=	2.0
3	7.0	—	1.0	=	6.0
4	8.0	—	6.0	=	2.0
5	10.0	—	4.0	=	6.0
6	15.0	—	13.0	=	2.0

This concludes our treatment of networking principles. What now remains is for us to illustrate the actual use of these principles. We shall observe how managers, by using these few simple ideas, can more effectively plan, see trouble spots in the future, and cope with them in an orderly fashion.

6

Network Replanning
and Adjustment

When the work breakdown schedule has been completed, when the network has been drawn, when all the time estimates for the expected elapsed times of the activities have been calculated, when all the T_L's and T_E's have been figured, and finally when the critical path has been identified—at this point when we might think that our work was over—in reality *it has just begun.*

PERT cannot be considered a sterile process of calculating times, drawing networks, and figuring slack time values; rather, it is a dynamic process involved with change, with readjustment, with the formulation of new networks when there are changes in schedules, and with constant revision of plans to achieve better performance in the light of changing conditions. For this reason the process of readjusting and replanning a PERT network is of prime importance to us. Three methods are shown.

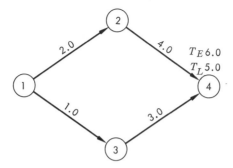

Figure 6–1. Network to be used for replanning.

Interchanging resources

To introduce this highly important and vital subject and to illustrate how the subjects of replanning and adjustment come to light early in the PERT process, let us look at the network illustrated in Figure 6–1. Further, let us suppose that each of the four activities represents work which is entirely manual, done entirely by unskilled labor—digging a group of four identical ditches, for instance. The first feature we notice about this network is that the

Figure 6–2. Figure 6–1 with the earliest expected dates and latest allowable dates added.

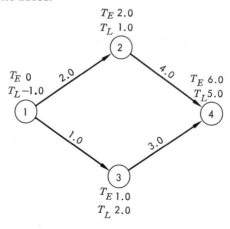

T_L is less than the T_E for the network-ending event. The way the network is laid out now indicates that 6.0 weeks are needed for completion, but because of some other schedule, some other connecting or coordinating project, it is apparent that the T_L, our scheduled completion date, must be 5.0 weeks, 1.0 week sooner than the earliest expected date of event 4.

A first approach might be to calculate whether slack time is present for any of the network events. If such is the case, then we have some maneuvering room or time to spare and can possibly shorten the time required to complete the project. Using methods outlined in Chapter 5, we have calculated T_E, T_L, and slack for each event in the network (Figure 6–2) and presented the results in Table 6–1.

At this point we are obviously interested only in positive slack. As we pointed out earlier, the connotation of negative slack is that we are behind and this of course we know already. We have been given 1.0 week less than we figured it would take to do the work. We find by analyzing Figure 6–2 and Table 6–1 that event 3 has positive slack time amounting to 1.0 week. This means that activity 1-3 could be 1.0 week late and still not interfere with the project's being finished in 5.0 weeks, the desired time.

You will remember that we said all four activities concerned themselves entirely with digging identical ditches by manual labor. Suppose that four men are working on activity 1-3 and two men on activity 1-2. Suppose further that we remove half the men who are working on activity 1-3. Then activity 1-3 will take twice as long, 2.0 weeks. Let us take the men we removed and put them to work on activity 1-2, one of the activities on the critical path.

Table 6–1. T_E, T_L, and Slack for Each Event in the Network in Figure 6–2

EVENT	T_L	—	T_E	=	SLACK
1	−1.0	—	0	=	−1.0
2	1.0	—	2.0	=	−1.0
3	2.0	—	1.0	=	1.0
4	5.0	—	6.0	=	−1.0

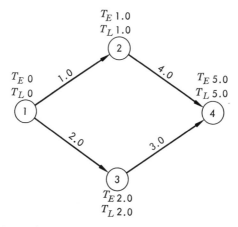

Figure 6–3. Figure 6–1 after simple replanning.

These extra men will make it possible for activity 1-2 to be completed in 1.0 week. How? If removing a number of men from activity 1-3 added 1.0 week to the work time required, then adding this same number of men to activity 1-2 will no doubt remove the same week from the time required there. Now that we have done this simple network replanning, let us draw the resulting new network and see what we have accomplished, as in Figure 6–3 and Table 6–2.

Analyzing the slack in Table 6–2 indicates that our replanning efforts have completely removed the slack time that existed, that there is no more slack in the system. What have we accomplished? Well, we have reduced the time required to perform this work to the acceptable level, 5.0 weeks, without adding any new resources, men in this case. Notice finally that when the network has been

Table 6–2. T_E, T_L, and Slack for Each Event in the Network in Figure 6–3

EVENT	T_L	—	T_E	=	SLACK
1	0	—	0	=	0
2	1.0	—	1.0	=	0
3	2.0	—	2.0	=	0
4	5.0	—	5.0	=	0

replanned so that no slack remains, all paths are critical paths; a slowdown in either of the two paths which remain will delay completion of the project beyond the 5.0 weeks allowed.

There are two lessons to be learned from the example we have just completed. One is that it was a trivial example, which we could have worked out in our heads without resorting to a lot of PERT methods, *but* when the networks become more complicated, as they will in later chapters, we shall not be able to find the right answer so easily. When the network is really complicated (say, more than 10 events), an orderly method of analysis such as that we just went through for our trivial problem will yield optimum results, whereas "playing it by ear" may not yield the optimum answer. The second lesson? In our example above, we interchanged resources which were *identical* man-hours of manual labor. If the work involved in activity 1-2, for instance, had been bricklaying and the work involved in activity 1-3, for instance, had been pipe fitting, then the laborers involved could not have been switched around quite so easily, probably not at all. In our case, however, we assumed that the work involved in all four activities in the network was of such a nature that any manual laborer could perform it just as well as any other. This assumption does not hold true for all types of projects, and thus the method of replanning just treated does not work in all cases.

Before we go on to another kind of replanning and adjustment technique, let us try replanning a slightly more complicated network, such as that illustrated in Figure 6–4. Let us assume once

Figure 6–4. Network to be used for replanning.

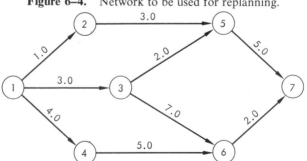

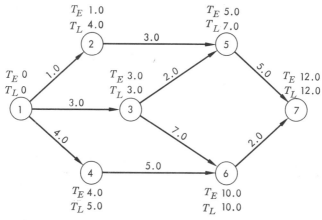

Figure 6–5. Figure 6–4 with the earliest expected dates and latest allowable dates added.

more that all the work involved in this project can be done by the same classification of labor and that such labor is freely interchangeable.

Our goal is the reduction of project time below 12.0 weeks. To begin the solution, calculate T_E and T_L for each of the events on the network and make another slack time listing as we did in the previous example. This has been done in Figure 6–5 and Table 6–3. We have arbitrarily set the T_L of the network-ending event at 12.0 weeks. As we shall show later, this really does not make any difference because we shall know when we have reached our minimum time solution.

Table 6–3. T_E, T_L, and Slack for Each Event in the Network in Figure 6–5

EVENT	T_L	—	T_E	=	SLACK
1	0	—	0	=	0
2	4.0	—	1.0	=	3.0
3	3.0	—	3.0	=	0
4	5.0	—	4.0	=	1.0
5	7.0	—	5.0	=	2.0
6	10.0	—	10.0	=	0
7	12.0	—	12.0	=	0

The purpose of calculating the slack time is to determine whether there is any slack in the system. If there is no slack, the chance of reducing the T_E of the network-ending event is slim _small_ indeed. Three of the events in this network have slack time, events 2, 4, and 5. The total slack time is 6.0 weeks, so that when we begin we know at least that the T_E of the network-ending event _can_ be reduced. There are four possible paths through the network:

1-2-5-7	9.0 weeks
1-3-5-7	10.0 weeks
1-3-6-7	12.0 weeks (critical path)
1-4-6-7	11.0 weeks

If all the inputs to this network are labor, the total man-weeks of work is then the total of the expected elapsed times of all the activities, or 32.0 weeks. The only significance of this figure is that we cannot accept as a legitimately readjusted network one in which the total expected elapsed times of all the activities add up to more than 32.0 weeks of labor inputs.

Now all that remains for us to do is shift labor around so that each of the four possible paths through the network requires exactly the same time. Many such solutions are possible; if we began this readjustment process before the actual work had begun, the number of possible ways to get all the paths equal is quite large. As work on the project progresses, however, and as some of the early activities (1-2, 1-3, and 1-4) are completed, any readjustment process must be carried out on the activities which remain, and the number of choices diminishes. If you do some

Figure 6–6.   Figure 6–4 after simple replanning.

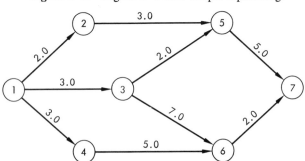

planning early, you are generally not so constrained as if you wait until later and then find fewer moves possible.

Rather than take the time and space to relate all the readjustments that are possible, we have elected to illustrate one of them. Let us start by shifting labor inputs on paths 1-2-5-7 and 1-4-6-7 until the two paths are equal. We have done this in Figure 6–6 simply by removing some labor from activity 1-2 and reapplying it to activity 1-4. Now we have four paths with total expected elapsed times as follows:

1-2-5-7	10.0 weeks
1-3-5-7	10.0 weeks
1-3-6-7	12.0 weeks
1-4-6-7	10.0 weeks

The second step in readjustment will be to look at the new paths 1-3-5-7 and 1-3-6-7. Suppose that we remove one unit of labor from activity 3-5 and add it to activity 3-6. Our results are illustrated in Figure 6–7. The expected elapsed times of the four possible paths are as follows:

1-2-5-7	10.0 weeks
1-3-5-7	11.0 weeks
1-3-6-7	11.0 weeks
1-4-6-7	10.0 weeks

Our final step will be to remove ½ unit of labor from each of activities 2-5 and 4-6 and add them, respectively, to activities

Figure 6–7.   Figure 6–6 after replanning.

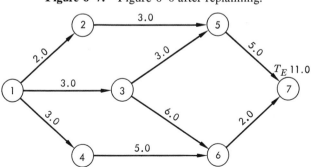

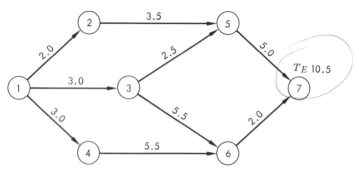

Figure 6–8. Figure 6–7 after replanning.

3-5 and 3-6. This will balance up the four paths in the network and yield the result shown in Figure 6–8.

1-2-5-7	10.5 weeks
1-3-5-7	10.5 weeks
1-3-6-7	10.5 weeks
1-4-6-7	10.5 weeks

One final word of caution before we go on to another method of network revision: We have been talking about shifting labor around from one activity to another. As long as the type of labor is identical or even quite similar, this is no problem, as we have seen. The rate, however, at which labor on one activity can substitute for labor on another activity is not always linear. For instance, take the two activities illustrated in Figure 6–9 with their respective expected elapsed times. Assume that the same number of men is assigned to each activity. If we wanted to shift some of the labor from activity 1-2 to activity 1-3, we could do so;

Figure 6–9. Network used to illustrate shifting resources.

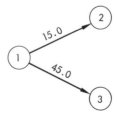

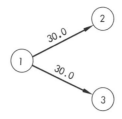

Figure 6–10. Figure 6–9 after replanning.

suppose we start by shifting half the labor from activity 1-2 to 1-3. The resulting partial network would be that illustrated in Figure 6–10.

Now imagine that we shift *all* the labor from activity 1-2 to activity 1-3. Does the expected elapsed time for activity 1-3 go to zero? It certainly does not. Shifting all the labor in the world to activity 1-3 could not reduce its time to zero, but we must understand why. Visualize a ditch in which 2 men are working, and assume that the work will be finished this afternoon sometime. If 2 more men were added to the work force, we would expect to finish in about half the time. But add 100 more men and watch what happens to the work. They would be so much in one another's way that probably the ditch would take longer than originally expected with only 2 men. This is referred to as the principle of diminishing returns. It shows why the transfer of additional labor to an activity cannot improve production indefinitely. The degree to which labor (or *any* resource) can be switched from one activity to another with proportional results in time reduction depends of course upon the peculiar nature of each individual project. There are no hard-and-fast rules that govern this type of substitution. Use of overtime is synonymous with increasing the labor input to an activity.

Relaxing the technical specifications

A second method of reducing the time required to complete a certain project would be to relax some of the technical specifications governing the project. For instance, if one of our technical requirements for a certain project is that paint must be allowed

Figure 6–11. Painting a building.

to dry for .4 week between coats, and we want to put two coats of paint on a certain building, our network might look something like that in Figure 6–11.

Activity 1-2 represents putting on the first coat, activity 2-3 represents the drying time, and activity 3-4 represents putting on the final coat. If we relax the technical specifications somewhat, perhaps to .3 week or about 2 days between coats, we can obviously reduce the T_E of the network-ending event. The extent to which this can be done is severely restricted in many cases. Take for instance the process of pouring concrete. If the specifications call for concrete to cure (a process that is often called setup) for 5 days before a load is put on it and we arbitrarily reduce this specification to 2 days, we may experience disastrous results when the finished concrete is put under load.

Another example of relaxing or reducing the technical specifications on a particular activity would be reduction of testing. Suppose that a part of a network involving the development of an aircraft pump has to do with testing. The original specifications state that the pump is to be tested under conditions of extreme cold, extreme heat, extreme change from heat to cold, and finally, extreme change from cold to heat. Common sense might dictate that this series of tests may have redundancy in it, and that one test or more could be eliminated. Specifications could be reduced without critical results, but with considerable saving of time in the network.

Changing the arrangement of activities

Considerable savings of time often can be effected by re-arranging the structure of the activities in the network. Let us first look at a simple example from production control and then see how the principle illustrated can be applied in the rearrangement and rescheduling of certain PERT networks.

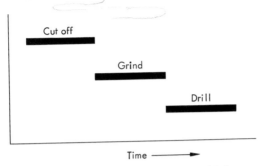

Figure 6–12. Series-connected machining activities for batches of 100 pieces.

Suppose that a particular finished part must go through three machining operations before completion. If these parts were sent through the machine shop in batches of 100 and if we wanted all pieces to travel along together, we would represent the process on a chart such as Figure 6–12. Activities represented in this manner are called *series-connected* in PERT terminology, meaning that one must be completed before the next can be begun. Certain parts of many projects are series-connected. Take for instance the two related activities of sanding the hull of a boat and then painting it. If we try to represent these as anything but series-connected we are in trouble right away, as anyone who has done any painting knows well. We would be forced to finish sanding, clean off the dust, and then paint in dirt-free conditions for any acceptable finish at all.

But back to our example. When a few parts have gone through the cutoff process, what would be wrong with sending them along to the grinder instead of waiting for the entire batch of 100 to be processed at the cutoff station? When a few units have gone through the grinding process, what would be wrong with sending them along to the drill presses instead of waiting until the entire batch has been processed? The rearranged activities portrayed in Figure 6–13 reflect a considerable saving in time.

This maneuver is well known to individuals who have worked in the area of production control. In PERT terminology, we have changed series-connected activities to series-parallel activities; this

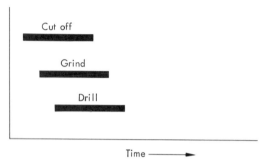

Figure 6–13. Figure 6–12 rearranged into series-parallel activities.

means that they now go on concurrently. Notice that we said *concurrently* and not *simultaneously*. It would be impossible for the first unit in our machine shop example to be cut off, ground, and drilled simultaneously; series-parallel activities are ones that can go on concurrently; different activities can be in operation at the same time.

Changing series-connected activities into series-parallel activities is thus one method of reducing the time required to complete a project. Expressing the idea with the simple charts in Figures 6–12 and 6–13 was easy, but how can this be translated into acceptable PERT notation? Suppose we use the same example— a part which must go through three manufacturing processes for completion. Let us assign identical times to each of these processes for the batch of 100 pieces, as follows:

Cutoff	1.2 weeks
Grind	1.2 weeks
Drill	1.2 weeks

In Figure 6–14 these three activities are represented as three series-connected activities in PERT network form. The T_E of the network-ending event (drilling completed) would be the sum of

Figure 6–14. Figure 6–12 in network form.

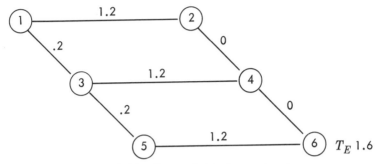

Figure 6–15. Figure 6–14 after replanning.

the expected elapsed times of the three activities, or 3.6 weeks. Now suppose that we wanted to change this to a series-parallel arrangement. We first of all have to decide how much time should elapse between the time when we started the cutoff process and the time when we began grinding. Suppose we let this magnitude be .2 week. We then assign the same .2-week time value to the time that will elapse between the start of the grinding operation and the start of the drilling operation. The resulting network is represented in Figure 6–15. Table 6–4 explains some of the activities in order to clarify just what we have done.

Table 6–4. Explanation of Activities in Figure 6–15

ACTIVITY	EXPLANATION
1-2	100 pieces are cut off.
1-3	.2 week elapses between start of cutoff activity (1-2) and the time sufficient pieces have been readied for the start of activity 3-4 (grind).
2-4	Dummy or zero-time activity to maintain network appearance.
3-4	100 pieces are ground.
3-5	.2 week elapses between start of grinding activity (3-4) and the time sufficient pieces have been readied for the start of activity 5-6 (drilling).
5-6	100 pieces are drilled.
4-6	Dummy or zero-time activity to maintain network appearance.

Rearrangement of our activities from series-connected to series-parallel has allowed us to reduce the T_E of the network-ending event from 3.6 weeks to 1.6 weeks. Of course, the amount of the reduction was determined by the arbitrary time periods we assigned to activities 1-3 and 3-5, the activities that represented waiting time between the start of one activity and the accumulation of enough partially processed items to justify the start of the next activity. Had we assigned larger time values to these two activities, the T_E of the network-ending event would have been increased accordingly.

There are many activities which cannot be handled in this manner. We have noted one activity which must be completed before its successor activity can begin at all: sanding must be finished before painting begins. Another example is the case in which a complete item must be produced before any sort of testing commences. In that case, production of the item and testing of the item could hardly go on concurrently.

There are many possibilities for actually readjusting or rearranging a PERT network; some networks have as many as several thousand activities. In a complicated network, advance determination of just how the critical path will be affected, just what the new critical path will be, or just how much of a reduction we can make in the T_E of the network-ending event can be quite difficult. Consider for a moment how difficult the rearrangement example first introduced in Figure 6–4 would have been had the network contained 100 rather than 7 events. For this reason, larger networks demand the use of electronic data-processing equipment; we have therefore included Chapter 8 on this topic.

7

Probability Concepts

U_p to this point we have been working with PERT
to estimate how long a project may take, whether it
can indeed be completed by its scheduled completion
date, and problems of this sort. All this work contains a certain amount of uncertainty because we are
not exactly sure that the answers to our mathematical
exercises are correct. The reason for this is simply
that our whole PERT system is based, you will recall,
upon estimates of time—expected elapsed time for
an activity, for instance. In Chapter 4 we learned
that three time estimates are required to calculate t_e:
(1) the pessimistic time, (2) the most likely time,
and (3) the optimistic time. These estimates involved
some uncertainty because the estimator was not positive that the job would be done in the most likely
time, nor was he positive that it would take as long
as the most pessimistic time, or that it could be

finished in the most optimistic time. Rather, we were told to state all three of these times and then calculate t_e by using the formula

$$t_e = \frac{a + 4m + b}{6}$$

Because the estimates of a, m, and b themselves involve uncertainty, there must of course be some uncertainty in the final answer for t_e and, following this to its logical conclusion, some uncertainty about all the time values in any network. This chapter deals with methods for treating these uncertainties by the use of statistics and probability theory. From the operational point of view, this will allow a manager not only to state that the T_E of the network ending event is 13.0 weeks, but also to state the odds, chances, or probability that the network will indeed be finished in 13.0 weeks.

For a quick example of the value of this type of statistical treatment, two earliest expected dates with their corresponding odds for completion are indicated in Table 7–1.

From the manager's point of view, the addition of probability values (which are nothing more than estimates of odds or chances) to the T_E values makes him better able to make decisions. For instance, in network 1 of Table 7–1 if there is a strong need to have the network completed by the end of 10.0 weeks, the manager should make some adjustments or take some action. Chances of 4 out of 10 are not very good odds that he will finish within the required time. On the other hand, looking at the second network below, if the work is needed to be completed within 12.0 weeks, the manager can feel fairly secure that he will hit his target since the calculated chances of finishing on time (that is, within 12.0 weeks) are high—9 chances out of 10. The addition of these probability values to the network time considerations allows the manager to have a more complete picture of the chances of com-

Table 7–1. Two Earliest Expected Dates with Odds for Completion

NETWORK	T_E, WEEKS	PROBABILITY THAT IT WILL BE COMPLETED BY THIS DATE
1	10.0	.40
2	12.0	.90

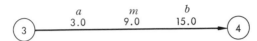

Figure 7–1.   Three estimates of expected elapsed time.

pleting the network. This chapter will be devoted to methods by which we can calculate and use these probability values.

In the original calculation of expected elapsed time for each activity, three time estimates were given. They are illustrated in Figure 7–1. The meaning and significance of Figure 7–1 are that an estimator familiar with the nature of the job to be completed in this particular activity has estimated that, at worst, it will take 15.0 weeks to do the work, that at best it could be done in 3.0 weeks, and that the most likely time is about 9.0 weeks. The very fact that he has given three different estimates shows that he really does not know (and cannot hope to know) the exact time the activity will take.

Recall from Chapter 4 that the *a* value, the most optimistic time, is the value below which there is only 1 chance in 100 the time could be. In this case the estimator is saying to us that 3.0 weeks is the quickest time we could hope to complete the work in —there is only 1 chance in 100 that the project could be done in less time. On a bell-shaped curve our estimator has given us a time estimate that belongs at the extreme left-hand end of the curve, as shown in Figure 7–2.

Figure 7–2.   Bell-shaped curve showing the most optimistic time of completion.

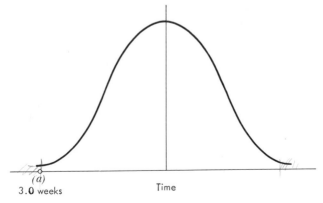

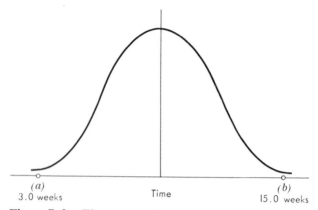

Figure 7–3. Figure 7–2 with the most pessimistic time of completion added.

On the other hand, when our estimator gives us the most pessimistic time value *b,* in this case 15.0 weeks, he is giving a time value which in his mind represents the time beyond which there is only 1 chance in 100 the network could require for completion. He has thus given us the time value for the extreme right-hand portion of the curve, as shown in Figure 7–3.

Of course, there is still a chance that the project could be completed in less than 3.0 weeks, and still a chance that it could require more than 15.0 weeks, but both of these chances are small. We illustrate these small chances by showing the *a* and *b* values at the outside points of the curve, but we leave a little area under the curve beyond these two points, recognizing the slight possibility for developments to be better than *a* or worse than *b*. These two areas are shaded in Figure 7–4. Compared to the entire area under the curve, these are miniscule indeed. The area under the curve between *R* and *S* represents the very small chance that the project could be completed in less than 3.0 weeks; similarly, the very small area under the curve between *T* and *U* represents the very small chance that the project will require more than 15.0 weeks for completion. Logically, then, we can say that the very large portion of the area under the curve between *S* and *T* represents the chance that the project will be completed within 3.0 to 15.0 weeks.

In Chapter 4 we introduced the concept of the standard deviation. We said it was a measure of the tendency of the curve to

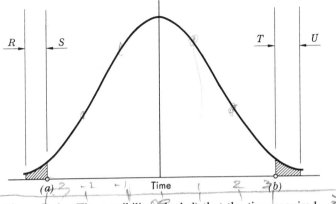

Figure 7–4.   The possibility (shaded) that the time required for a job will be less or more than the expected minimum and maximum.

spread out or of the values to disperse. We illustrated the manner by which it is calculated and made several statements about its properties, which are repeated below.

There is mathematical proof that:

1. Approximately 68 percent of all the values in a bell-shaped distribution lie within ±1 standard deviation from the average
2. Approximately 95 percent of all the values lie within ±2 standard deviations from the average
3. Approximately 99.7 percent of all the values lie within ±3 standard deviations from the average

The third of these facts is the one which allows us to calculate the standard deviation for our activities quite easily. If 99.7 percent of all the values (all the possible completion times in this case) do lie between 3.0 and 15.0 weeks, then the distance from the extreme left-hand end (3.0 weeks) to the extreme right-hand end of the curve (15.0 weeks) must be ±3 standard deviations, or 6 standard deviations in total. As this is true, then 1 standard deviation for the activity first shown in Figure 7–1 must be calculated by:

$$\frac{15.00 - 3.0}{6} = 2.0 \text{ weeks}$$

In a sense, what we have said is simply that if the estimator feels that his range of 3.0 to 15.0 weeks includes about all the possible values under the curve, he is really saying that the distance between these two estimates is about 6 standard deviations. This method allows us to compute the standard deviation values easily and with reasonable accuracy. It should be clear that we could not use the method first introduced in Chapter 4, because that required observation of the event many times and then the calculation of the standard deviation value. In PERT the emphasis, of course, is on one-time, nonrepetitive projects for which there is no history of the activity; we must do the best we can by using the time estimates given by the project estimator. There is mathematical basis for assuming that the value for the standard deviation calculated in this manner is accurate enough for our purposes.

Suppose we illustrate the estimates for three activities (Figure 7–5) and calculate the standard deviation for each of these as follows:

Activity standard deviation:

$$\frac{7.0 - 4.0}{6} = .5$$

$$\frac{10.0 - 1.0}{6} = 1.5$$

$$\frac{21.0 - 3.0}{6} = 3.0$$

Notice that in all cases the value of the most likely time m in no way affects the calculation of the standard deviation. The standard deviation is affected only by the relative distance from the most optimistic estimate to the most pessimistic estimate. Indeed, one can notice from Figure 7–5 and the accompanying calculations that the values for m vary between the two other estimates. Notice particularly in example 1 that the most likely time estimate is nearer to b than it is to a; notice the opposite relationship in example 2: m is quite a bit nearer to the a value than it is to the b value; finally, notice in example 3 that the m value is exactly midway between the most optimistic value a and the most pessimistic value b. In example 1 the estimator is telling us that it is more

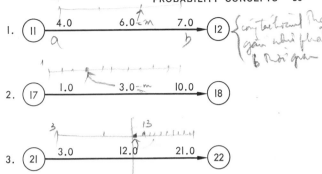

Figure 7–5. Time estimates for three activities.

likely that the project will require the full 7.0 weeks than that it will be finished in just 4.0 weeks; thus, we find the most likely estimate nearer to the most pessimistic estimate. In example 2, on the other hand, the estimator is telling us that he is somewhat surer that he will finish early than that he will finish late; the most likely time estimate is nearer to the most optimistic time than it is to the most pessimistic time. In example 3, the estimator is about as certain that the project will run over the most likely time as he is that it will be completed in less than the most likely time; thus, the *m* value is exactly halfway between the *a* and *b* values. As we have already pointed out, the standard deviation is a measure of the tendency of data to spread out. Thus, we are interested only in the values for the most optimistic and most pessimistic times, as these indicate the extreme right- and left-hand ends of the range of values, and *m* does not enter the calculation.

Let us now consider several series-connected activities such as those shown in Figure 7–6. Values for *a, m,* and *b* are given for each of the activities. First we shall calculate the standard deviation for each of the individual activities, and then we shall illustrate how this task is performed for a group of connected activities. Calculation of individual standard deviations is shown in Table

Figure 7–6. Series-connected activities with three time estimates.

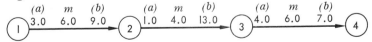

Table 7–2. Calculation of Individual Standard Deviations for Activities in Figure 7–6

ACTIVITY	a	b	$b - a$	ACTIVITY STANDARD DEVIATIONS $\dfrac{b - a}{6}$
1-2	3.0	9.0	6.0	1.0
2-3	1.0	13.0	12.0	2.0
3-4	4.0	7.0	3.0	.5

7–2. From these calculations, we have measures of the dispersion (spread) of the activities around their most likely times. In other words, we have an idea of how certain the three estimators were about their estimates.

The next step is to calculate the expected elapsed time for each of the activities and the earliest expected date T_E for the network-ending event, event 4. This is done in Table 7–3.

The next step is to add together the expected elapsed times for the activities to find the earliest expected date T_E for the network-ending event. Because the activities were arranged in series-connected form, the job involves only adding them together. If the network were of a more complicated form, we would have to observe the networking rules laid down in Chapter 5.

$$T_E \text{ (event 4)} = 6.0 + 5.0 + 5.8$$
$$= 16.8 \text{ weeks}$$

Table 7–3. Calculation of Expected Elapsed Times and Earliest Expected Date for the Network-ending Event in Figure 7–6

ACTIVITY	a	b	m	$\dfrac{a + 4m + b}{6}$	t_e
1-2	3.0	9.0	6.0	$\dfrac{3.0 + 24.0 + 9.0}{6}$ =	6.0
2-3	1.0	13.0	4.0	$\dfrac{1.0 + 16.0 + 13.0}{6}$ =	5.0
3-4	4.0	7.0	6.0	$\dfrac{4.0 + 24.0 + 7.0}{6}$ =	5.8

To calculate a probability measure that will help us know what our chances are of finishing on time, let us start by calculating the standard deviation for the ending event of this series, event 4. To calculate the standard deviation for the ending event of several activities connected in series, take the square root of the sum of each of the individual activity standard deviations squared. This sounds more formidable than it really is.

Standard deviation for event 4

$$= \sqrt{\left(\begin{array}{c}\text{Std. deviation} \\ \text{activity 1-2}\end{array}\right)^2 + \left(\begin{array}{c}\text{Std. deviation} \\ \text{activity 2-3}\end{array}\right)^2 + \left(\begin{array}{c}\text{Std. deviation} \\ \text{activity 3-4}\end{array}\right)^2}$$

$$= \sqrt{(1.0)^2 + (2.0)^2 + (.5)^2}$$

$$= \sqrt{1 + 4 + .25}$$

$$= \sqrt{5.25}$$

$$= \text{approximately 2.3 weeks}$$

Now we have two measures of this network:

1. Its T_E
2. The standard deviation of the network-ending event—symbolized σT_E

These two values are 16.8 and 2.3 weeks, respectively.

Let us try a few more of these and then illustrate how this new bit of information will be of value to us in managerial decision making about networks. Following are three networks, with *a, b,* and *m* values shown for each activity (Figures 7–7 to 7–9). We shall calculate:

1. Expected elapsed time for each activity
2. Standard deviation for each activity
3. T_E—earliest expected date for the network-ending event
4. σT_E—standard deviation for the network-ending event

Note that there is only one path through the network; therefore it must be the critical path.

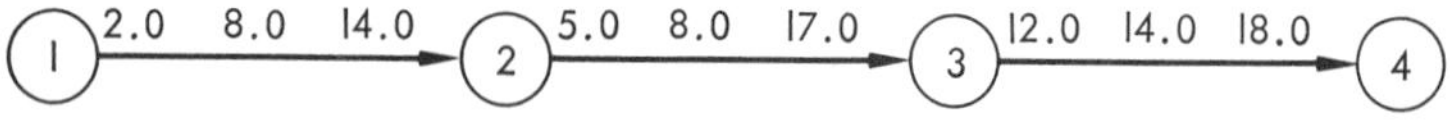

Figure 7–7. Network with three time estimates shown.

Calculations for network in Figure 7–7

Activity 1-2

$$t_e\text{'s} = \frac{2.0 + 32.0 + 14.0}{6}$$

$$= \frac{48.0}{6}$$

$$= 8.0 \text{ weeks}$$

Activity 2-3

$$= \frac{5.0 + 32.0 + 17.0}{6}$$

$$= \frac{54.0}{6}$$

$$= 9.0 \text{ weeks}$$

Activity 3-4

$$= \frac{12.0 + 56.0 + 18.0}{6}$$

$$= \frac{86.0}{6}$$

$$= 14.3 \text{ weeks}$$

Activity standard deviations =	Activity 1-2	Activity 2-3	Activity 3-4
	$\frac{14.0 - 2.0}{6}$	$\frac{17.0 - 5.0}{6}$	$\frac{18.0 - 12.0}{6}$
	= 2.0 weeks	2.0 weeks	1.0 week

Standard deviation for event 4 $= \sqrt{(2.0)^2 + (2.0)^2 + (1.0)^2}$

$$= \sqrt{9.0}$$

$$= 3.0 \text{ weeks}$$

T_E for event 4 $= 8.0 + 9.0 + 14.3$

$$= 31.3 \text{ weeks}$$

Event

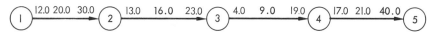

Figure 7–8. Network with three time estimates shown.

Calculations for network in Figure 7–8

Activity 1-2

$$t_e\text{'s} = \frac{12.0 + 80.0 + 30.0}{6}$$

$$= \frac{122.0}{6}$$

$$= 20.3 \text{ weeks}$$

Activity 2-3

$$= \frac{13.0 + 64.0 + 23.0}{6}$$

$$= \frac{100.0}{6}$$

$$= 16.7 \text{ weeks}$$

Activity 3-4

$$= \frac{4.0 + 36.0 + 19.0}{6}$$

$$= \frac{59.0}{6}$$

$$= 9.8 \text{ weeks}$$

Activity 4-5

$$= \frac{17.0 + 84.0 + 40.0}{6}$$

$$= \frac{141.0}{6}$$

$$= 23.5 \text{ weeks}$$

	Activity 1-2	Activity 2-3
Activity standard deviations =	$\dfrac{30.0 - 12.0}{6}$	$\dfrac{23.0 - 13.0}{6}$
	= 3.0 weeks	1.7 weeks

$$\begin{array}{cc} & \text{Activity 3-4} & \text{Activity 4-5} \\ = & \dfrac{19.0 - 4.0}{6} & \dfrac{40.0 - 17.0}{6} \\ = & 2.5 \text{ weeks} & 3.8 \text{ weeks} \end{array}$$

Standard deviation for event 4

$= \sqrt{(3.0)^2 + (1.7)^2 + (2.5)^2 + (3.8)^2}$

$= \sqrt{32.6}$

$= 5.7$ weeks

T_E for event $4 = 20.3 + 16.7 + 9.8 + 23.5$

$\qquad = 70.3$ weeks

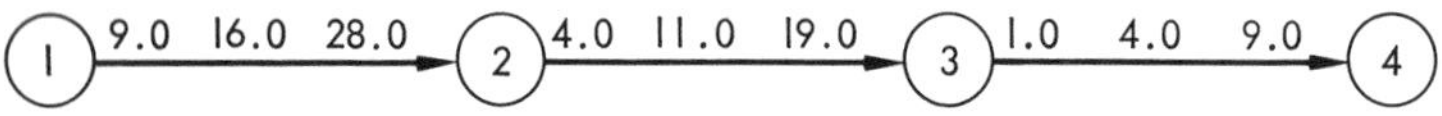

Figure 7–9. Network with three time estimates shown.

In addition to knowing the standard deviation for the ending event of each of these networks, we might at some time like to have the standard deviation for an event *other* than the ending event. For instance, in the network in Figure 7–9, management might want to know the chances of completing an event other than the network-ending event by its latest allowable date. This might give management an idea of whether or not the probability of completing event 3, for example, on time is great enough to obviate further thought or, perhaps, further shifting of resources. To calculate the standard deviation of event 3 in Figure 7–9, it is necessary to work only with the activities up to that event, activities 1-2 and 1-3. Because activity 3-4 comes after the event in question, it cannot affect event 3. Calculation of the T_E's and standard deviation for event 3 in Figure 7–9 is as follows:

Calculations for network in Figure 7–9

Activity 1-2

$$t_e\text{'s} = \dfrac{9.0 + 64.0 + 28.0}{6}$$

$$= \dfrac{101.0}{6}$$

$$= 16.8 \text{ weeks}$$

Activity 2-3

$$= \frac{4.0 + 44.0 + 19.0}{6}$$

$$= \frac{67.0}{6}$$

$$= 11.1 \text{ weeks}$$

	Activity 1-2	Activity 2-3
Activity standard deviations $=$	$\dfrac{28.0 - 9.0}{6}$	$\dfrac{19.0 - 4.0}{6}$
	$= \quad 3.2$ weeks	2.5 weeks

Standard deviation for event 3 $= \sqrt{(3.2)^2 + (2.5)^2}$

$$= \sqrt{16.5}$$

$$= 4.1 \text{ weeks}$$

T_E for event 3 $= 16.8 + 11.1$

$$= 27.9 \text{ weeks}$$

We have now learned how to calculate the standard deviation for any event. How does this aid in managerial planning? Taken together, the T_E and the standard deviation of an event form a probability curve. If the standard deviation is *large* when compared with its T_E, variance is wide; the estimators whose estimates comprise this path are collectively not very certain about their estimates; the actual time could vary far from the T_E. If the standard deviation is *small* when compared with its T_E, the estimators are quite certain that the actual time will not be far from the T_E. To further illustrate this point, we have shown two probability curves in Figure 7–10. Figure 7–10a has a narrow spread—a small standard deviation compared to the T_E. Figure 7–10b has a wide spread—a large standard deviation compared to the T_E. Each curve contains a message for decision makers. The curve in Figure 7–10a implies that even if developments go badly and the schedule is not made, the amount of time actually used will not be much greater than the T_E value.

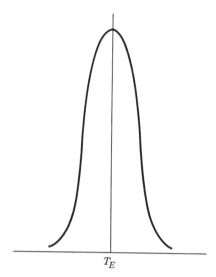

T_E

Figure 7–10a. Probability distribution with a narrow spread, or a small standard deviation compared with the earliest expected date.

On the other hand, however, if things go wrong in the case illustrated in Figure 7–10b, the amount of time used can go quite high, far to the right of the T_E value. Naturally, we always want to be involved in situations such as the one illustrated in Figure 7–10a so that our mistakes will not be too costly; unfortunately, we cannot always have what we want.

Figure 7–10b. Probability distribution with a wide spread, or a large standard deviation compared with the earliest expected date.

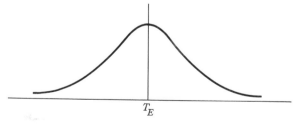

T_E

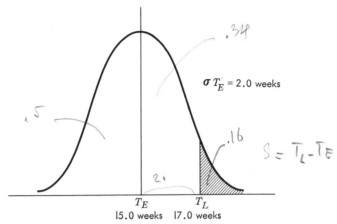

.34

$\sigma T_E = 2.0$ weeks

.16

$S = T_L - T_E$

.5

2.

T_E T_L
15.0 weeks   17.0 weeks

Figure 7–11. Probability distribution; shaded area illustrates probability of being late.

Let us illustrate how we can calculate our chances of completing a network by its latest allowable date T_L. A probability curve is illustrated in Figure 7–11, and the T_E and T_L of the network-ending event are noted, along with the standard deviation of that event.

Notice first that the chances of the project's requiring either more or fewer than 15.0 weeks are exactly .5 or ½. This is because half the area under the curve lies to the right of the T_E point and half to the left. But we need to know what the chances are of finishing before the latest allowable date T_L. The T_L lies 2.0 weeks to the right of the T_E value. Because 2.0 weeks is exactly 1 standard deviation, T_L lies 1 standard deviation to the right of T_E. From the list on page 97, we know that approximately 68 percent of all the values in a bell-shaped distribution lie within ±1 standard deviation from the average; thus, since the T_L point is exactly 1 standard deviation to the right (+1 standard deviation), about 34.0 percent of the values must lie between T_E and T_L. Because half the values under the curve lie to the left of T_E *and* because 34.0 percent lie between T_E and T_L, a total of 84.0 percent of the values lie between the left-hand tail and T_L. Thus, 83.5 percent (84.0 chances out of 100) is our probability or odds of finishing before the latest allowable date. Conversely, the difference between 84.0 and 100.0 percent (16.0 chances out of 100) is the probability or odds of the project taking longer than the T_L value. The shaded area illustrates the chances that we shall be late.

Now let us look at an example in which the T_L value lies to the left of the T_E value, that is, where we know that the odds are stacked against us from the beginning—that we have less time available to complete the job than we know we shall need—and figure the chances that we shall actually be late. Such a situation is shown in Figure 7–12, in which we see that T_L lies 2.0 weeks or 2 standard deviations to the left of the T_E value. From our study of the bell-shaped or normal curve we know that approximately 95 percent of all the values under the curve lie within ±2 standard deviations from the average. As T_L is exactly −2.0 standard deviations from T_E, the area under the curve between T_L and T_E must be ½ of 95, or 47.5 percent. If we add to 47.5 percent the 50.0 percent of the area lying between T_E and the right-hand tail of the curve, we get a total of 97.5 percent. This is the area under the curve from the T_L value to the right-hand end and indicates the chances of being late, 97.5 out of 100 in

Figure 7–12. Probability distribution; shaded area illustrates probability of being late.

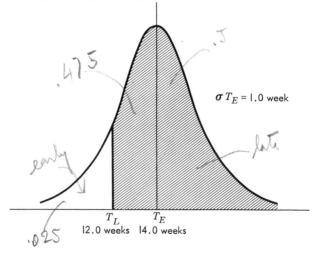

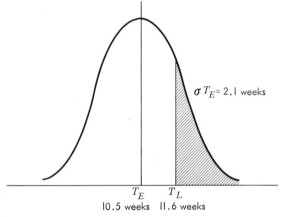

Figure 7–13. Probability distribution; shaded area illustrates probability of being late.

this case. With chances like this we are in trouble; early corrective action must of course be taken to reduce the earliest expected date by network revision.

But suppose that the standard deviation is not an even number and that the distance between T_L and T_E is not an even number of weeks. How can we calculate the probability of finishing on time? Figure 7–13 shows such a case. We can subtract the T_L from the T_E value and get the distance between them measured in weeks. You will recall that this distance is actually called slack and is designated S. The formula is

$$T_L - T_E = \text{slack}$$

Thus, the slack time on this particular critical path is

$$11.6 \text{ weeks} - 10.5 \text{ weeks} = 1.1 \text{ weeks}$$

If we now divide the slack time by the standard deviation of the network-ending event, we get a measure of the slack time in terms of the number of standard deviations it represents:

$$\frac{1.1}{2.1} = .52 \text{ standard deviation}$$

But how do we treat .52 standard deviation? Our experience up to this point has been only with 1, 2, or 3 standard deviations. Fortunately, there are tables which help us. Appendix IV, a table of areas under the curve, tells how many standard deviations are required to include any portion of the area under the bell-shaped curve measuring to the right from the left-hand end of the curve. In our particular case, we can look in the margin of the table and find that about .698 of the area under the curve is contained between the left-hand end and a point .52 standard deviation to the right of the average. Thus, chances are almost 70 out of 100 that we shall complete the job on time. Again, the shaded area (in Figure 7–13) represents the chances of not finishing by the T_L date, about 30 out of 100 in this case.

Let us take another case in which T_L lies to the left of the T_E value, where we are in trouble before we begin. Figure 7–14 illustrates this situation. Once again we subtract the T_E value from the T_L value:

$$T_L - T_E = \text{slack}$$
$$13.6 - 14.3 = -.7 \text{ week of slack}$$

The concept of negative slack is not new because it was treated in Chapter 5. It simply means that we are behind from the

Figure 7–14. Probability distribution; shaded area illustrates probability of finishing on time.

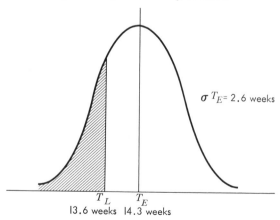

beginning. If we now divide this slack of $-.7$ week by the standard deviation of the network-ending event, we get

$$\frac{-.7}{2.6} = -.27 \text{ standard deviation}$$

Thus, we find that the T_L value lies $-.27$ standard deviation to the left of the T_E value. Now how do we assess the chances of being finished by the T_L time? Since the curve is symmetrical, as the two sides are identical, let us determine the area between the right-hand end and the T_L value. Looking in the margin of the table for the $-.7$ value we see that 75.8 percent of the area lies between T_L and the *right-hand* edge. Note that we are using the reverse method this time. This of course means that the balance of the area under the curve (24.2 percent) lies between the left-hand edge and the T_L value. The chances of finishing on time are therefore 24.2 per cent or slightly over 24 out of 100—not very good.

The ability to calculate the mathematical chances of finishing prior to the latest allowable date makes our decision making a little more precise; we no longer have to content ourselves with saying that we do not have enough time or that we do have enough time. We can now assess our condition by the use of probabilities and state what our chances are. Usually, in PERT, a probability of .6 or better of finishing on time is considered good.

8

Use of the Computer in PERT Applications

Today is the day of the computer. Its widest current application in industry is in the collection and processing of data in order to provide management with more complete information on which management can base decisions. It is quite natural, then, for computers to play a role in management's use of PERT.

When a network consists of very few activities, or when all its activities are personally known to and supervised by one manager, decisions concerning rearrangement, rescheduling, reduction, and such can be quite simple. However, development of the more complex networks, specialization of managerial functions, and an increase in the number of managerial echelons in a firm may lead to circumstances in which decisions must be based on secondhand data or information rather than on direct contact or personal observation.

When an organization becomes so large that persons making decisions in one part of the structure know little or nothing about what is going on in other parts of the structure, suboptimization may take place; that is, decisions may be made which the decision maker thinks are optimum when actually they serve to reduce the achievement of the total organization. Let us examine a PERT example: A unit manager alters the arrangement of two activities in his part of the project to save a few *hours* of time; the result is to set back the master project three to four *days*. The possibility of this happening causes intelligent managers to look toward electronic data-processing equipment as a possible aid in such situations.

Another reason why computers are being put to work in the PERT process is their ability to generate masses of information quickly. In many companies there are some who receive operating statements as long as two months after the close of the operating periods to which they pertain. Reports concerning what a manufacturer's product cost last month may not be available until after that product is actually in the hands of the ultimate consumer. Moving closer to PERT, considerable expense can be the result of management's lack of up-to-date knowledge about certain parts of a project. For instance, if management learns early enough that the steelwork in a large building under construction is behind schedule, some arrangement might be worked out for the bricklayers who are to follow the steelworkers. If this falling behind in steelwork becomes known only a day or two before the brickmasons are to report for work, probably the only outcome will be higher labor costs than budgeted for the project. In projects containing thousands of interconnected activities, information and data cannot be complete, accurate, and current without the use of electronic computers.

Computers can help in putting small networks together to form project networks or master networks. We indicated that the process of combining networks involves first the identification of interface events—events common to more than one network. For small networks, this identification is not difficult. For large, involved networks, the use of computers can help greatly in reducing the time necessary to prepare the master network.

Computers contribute to the more satisfactory presentation of data. A complete network consisting of 1,000 activities occupies so much space on paper as to be awkward and unwieldy to handle. Many persons are typically interested in and concerned with far less than the entire project network. A computer can digest the whole network, analyze it, pick out the salient information required by the various managers, and prepare and present those data in whatever variety of listings or output forms are desired.

Computers aid managers in discharging one of management's basic responsibilities: control. In order to be able to control, management must know at any time just where operations are in relation to where they were scheduled to be at that point in time. PERT functions both in setting up the most desirable schedule *and* in providing a method of checking actual production or progress against that schedule. This second job is essential if management is to keep posted, to stay abreast of developments and changes in hundreds—indeed thousands—of activities.

Imagine a manager telling his assistant, "Find out what slack time is available in the network this week," when the network contains 6,000 activities. If he gets data in time, the manager is in a position to rearrange the network, to shift resources, perhaps to reduce certain technical specifications in an attempt to complete the project in the shortest feasible time. Use of electronic data-processing machines is the *only* thing that permits adequate and accurate information to get to managers quickly.

When to use a computer

Before you have worked with PERT very long, you will be asking "How many activities must we have to justify the use of a computer?" Although this question has no concrete answer, we can at least make some helpful observations.

First, we must note the complexity of our network. When a network contains thousands of activities there is little doubt of the wisdom of using a computer in calculation and report generation. However, there are reported instances in which networks of 1,000 activities have been hand-computed. Such undertakings understandably require much time and many persons.

But stop and wonder for a moment how the project manager of such a network could get timely, up-to-date information about what was going on. Suppose during the course of just one week's work 20 of the 1,000 activities exceeded their time estimates. Would you like to have the task of calculating the new critical path so that the project manager could make some quick decisions? Of course not. Although sheer size of network does not demand a computer, the need for current information may recommend a computer's use.

Duration of the project and frequency of reports are factors to be considered. Unless the project goes exactly according to plans, and this is most rare, revision must take place each time an activity either exceeds its planned time or is completed in less than its planned time. This can call for frequent control reports, and on large involved networks, these reports may demand the use of a computer. Suppose, on the other hand, we estimate that a control report for a particular project should be prepared only once each two months. Clearly, if this project was originally scheduled for four months, we would question the wisdom of getting set up to use a computer for one control report. Thus duration of project and frequency of desired reports are both issues to be reckoned with in determining whether electronic data-processing equipment will be needed by the PERT user.

Types of computer outputs available

The total number of PERT reports that a computer could be programmed to prepare is limited only by the imagination of the individual programmer. There are, however, several general types of outputs that are useful to management personnel associated with PERT projects, as follows:

1. EVENT NUMBER REPORT

The event number report is the means of isolating and analyzing any particular event and the activities that lead to it. It tells how many and which activities lead to a particular event. Being able to analyze reported information on the activities that lead to a particular event enables management to understand how the event in question might be affected by its current and future position in the network.

The event number report also provides the earliest expected date for each of the ending events of each of the activities leading into the event. By analyzing this information, management can note which of the activities has the ending event with the latest expected date of completion and can thus be in a position to know which activity will be most likely to affect the accomplishment of the event in question.

The report provides the latest allowable date for each event. By having the latest allowable date available and by being in a position to observe the earliest expected dates of events leading up to the event in question, management is able to recognize probable critical spots in advance. In addition, when the latest allowable dates of events on the network are compared periodically with information concerning degree of completion and the current calendar date, assumptions can be made about the completion of the entire project.

Finally, the event number report makes available for each event the slack time associated with that event. The significance of slack time is, of course, that management, in replanning or readjusting the network or in reacting to unforeseen changes, can be properly apprised of the availability of surplus resources, the degree of criticality of each event, and where slippage in the network can most likely be made up. By analyzing successive reports, management can easily observe the movement of any events in question from the critical path to other noncritical paths and vice versa.

2. SLACK TIME REPORT

The slack time report lists each path through the network beginning with the critical path and going all the way down to the path with the greatest amount of slack time. The amount of slack associated with each path is so noted. Using this report, management can see at a glance the problems with each path; if the slack associated with a path is positive, the activities along that path can easily be delayed without increasing the earliest expected date of the network-ending event. Of course, the higher the positive slack value, the more delay it is possible to experience (or for re-planning purposes, the more resources it is possible to divert from that path to other more critical paths) without interfering with previously scheduled completion dates for the network. A negative slack indicates how much we must speed up completion of the activities on this path in order *not* to interfere with scheduled completion dates for the network. Replanning may in many cases take the form of reapplying resources from paths which contain positive slack to those which contain negative slack. If the network is rather large and no such slack time report is available, valuable decision-making time is lost while hand-computed solutions to the slack values are undertaken.

With modern computing machinery it is possible to generate slack value reports for various proposed solutions. The manager can choose several likely looking alternative solutions to a network problem involving, let us say, a network that is running behind schedule, then run these proposed solutions through the computer and have a slack time report printed out for each of the proposed solutions. Analysis of the reports indicates which of the proposed solutions comes nearest to alleviating the problem. This method is called simulation and represents perhaps the only way of forecasting what would happen if we did this or that without assuming the financial risk of actually trying out the proposals one by one in actual practice.

Like any other computational task, the slack time report could be hand-computed. To do so would, however, interfere drastically with the management decision-making process; while the actual work was going on at the project site, forward planning of this work would have to be stopped while hand computation of results was undertaken. Again, this would represent gross violation of the control principle: the feedback would so lag the actual completion of the work that readjustment of the network or replanning of actions would be impossible. Instead of allowing the control information to lag significantly behind the actual work completion, the use of a computer to generate the slack time report generally allows management to foresee problems that might otherwise not come to light until several months later and to take action then which will prevent these problems from actually happening. In that sense the slack time report is not only a control device but, in a measure, also an excellent method of intermediate-range planning.

3. LATEST ALLOWABLE DATE REPORT

The computer can be programmed to print out for the latest allowable date report the latest allowable dates for each of the events yet to occur in the network. Used in conjunction with the slack time report, this is a very effective planning tool. Comparison of latest allowable dates of events with current calendar dates is a method of determining one's current position; use of the slack time report is another.

4. DEPARTMENTAL REPORT

The departmental report is designed to provide planning and control information by department, enabling department heads to observe how *their* respective parts of the network are progressing. Once again, if the various department heads are given current performance information on each activity for which they are responsible, the project manager can legitimately expect a much better degree of coordination between departments and better overall project performance.

Let us repeat that these are only four types of reports. In many situations the information contained in one or more of these reports would not be sufficient for managerial decision-making purposes. In that case, the computer would be programmed to provide additional information needed by management. The extent to which this can be done is limited almost solely by the imagination of management and the technical ability of the programmer. Most computer manufacturers have developed "canned" PERT programs, i.e., programs that are already written; this considerably lessens the programming involved.

In general and in conclusion, management is interested in certain information that allows managers to plan, to direct, and to control effectively. In many cases the computer is able to yield this information more economically than hand computing. The types of information required, the time constraints imposed upon the generation and reporting of this information, the duration of the project in question, the availability of computation equipment, and the magnitude of the project all go together to determine the wisdom of using electronic data-processing equipment as an adjunct to the PERT process. The dominant consideration here is the same as in other management decisions—the effects on the long-range profit position of the firm.

9

Critical Path Method

Next to PERT, the critical path method (CPM) of planning and controlling projects has enjoyed the widest use among all the systems that follow the networking principle.

The fundamental departure of CPM from PERT is that CPM brings into the planning and control process the concept of *cost*. We do not mean to imply that PERT completely omits the cost concept; in PERT one must assume that *cost varies directly with time* for all activities within the project. Thus, when a reduction in time has been effected, we assume that a reduction in cost has also been achieved. We must, moreover, assume that a reduction of one week in time for one activity on the critical path is as productive of economy as any other reduction of one week in time on any other critical path activity. When the earliest expected date of the network-ending event has been reduced, there has been a reduction in cost.

A second major departure of CPM from PERT is found in the method of making time estimates. The CPM user is assumed to be on firmer ground when estimating the time required for the performance of each activity.

The major difference in the use of these two techniques can be grasped by contrasting one firm, a construction company, with another, a research and development company. An intelligent estimator for a construction company can give estimated cost and time figures for laying a concrete foundation. He may be off a little here and there, but if the company has done this type of work before, the estimates he makes about both cost and time will be acceptably accurate.

Now contrast this with the problem of the project director for a recently launched spacecraft. This project involves work never done before; the range of possible technical problems is immense; the number of subcontractors participating can run into the hundreds. Time estimates made for use in the research-type project may be little more than guesses.

As to the cost information in the PERT system, the project director might prefer to control *time* and let *costs* go their merry way. The job foreman for the construction company can hardly take the same attitude; he must complete the job on time and for the estimated cost.

We do not imply that CPM is exclusively for contractors and that PERT is useful only for developers of space programs. We do think those examples will indicate that differences in program needs will dictate the techniques used to plan and control the work to be done. When time can be estimated fairly well and when costs can be calculated in advance (labor and materials for a construction project, for example), CPM may represent the better of the two alternative control methods. On the other hand, when there is an extreme degree of uncertainty and when control over time outweighs control over costs, PERT or one of its variants may well represent the better choice.

The networking principles involved in CPM are like those covered in the PERT system. Thus, a person familiar with PERT has no trouble using CPM insofar as networking is concerned.

The first real departure is in time estimating, so let us get right to that point.

Under the CPM system, two time and cost estimates are indicated for each activity in the network; these two are a *normal* estimate and a *crash* estimate. The *normal* estimate of time approximates the most likely time estimate in PERT. Normal cost is, of course, the cost associated with finishing the project in the normal time. The *crash* time estimate is the time that would be required if no costs were spared in trying to reduce the project time. Under this program, the manager would do whatever was necessary to speed up the work. Crash cost, therefore, is the cost associated with doing the job on a crash basis in order to minimize completion time.

For example, let us look at a contracting situation involving road construction. Suppose that one activity is the grading of 10 miles of roadbed and that the normal time and cost figures are 3 months and $1 million. We know that this time estimate of 3 months could be reduced by the expenditure of more money, but just how? First, overtime work is possible—the working day can be increased from approximately 8 hours to approximately 12 hours. This would normally increase the direct labor cost 50 percent for all hours over 40 per week. Secondly, we might work 7 days a week instead of 5 or 6; if we assume a normal working agreement about wages and hours, this could increase direct labor cost by 100 percent for all hours over 48. Third, we might even put on a night shift for certain work. Although it would not be possible to do fine grading at night, we could haul in fill dirt, haul away dirt from cuts, and perhaps even do some rough grading at times. This would involve the addition of supervisory personnel for the second shift, some loss in efficiency from working under nonoptimum conditions, and finally much higher labor cost because of the difficulty of procuring workers willing to work at night. But the important fact is that these three *could all be done* if it were necessary to expedite the project. True, the cost would be high, but if those were the only ways to expedite the completion, then they would be distinct possibilities.

Suppose we represent our situation graphically, as in Figure 9–1. The vertical axis represents the cost of completing the project,

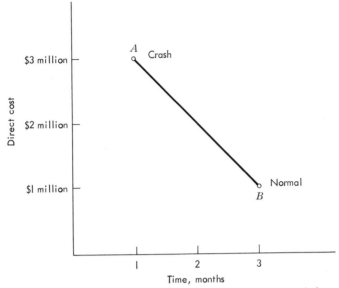

Figure 9–1. Crash time and cost compared with normal time and cost.

and the horizontal axis represents the time required for completion. We have the original normal estimates of 3 months and $1 million. Now suppose that with a crash effort we could complete the work in 1 month at a crash cost of $3 million. These two points are connected with a straight line in Figure 9–1. The line connecting points A and B in Figure 9–1 is referred to as the approximate time-cost curve. We say "approximate" because we do not know precisely how the time-cost relationship actually behaves without additional research and cost analysis on this activity. Figure 9–2 pictures a time-cost relationship to illustrate this point. In this case, the approximate curve is the solid line, and the true curve is the dotted line.

Figure 9–2 represents the case where an initial reduction of time can be effected with a modest increase in cost. Time required for the project can be reduced from point P to point O while cost rises only from point R to point S. This activity can be expedited, or its time for completion can be reduced significantly; initially,

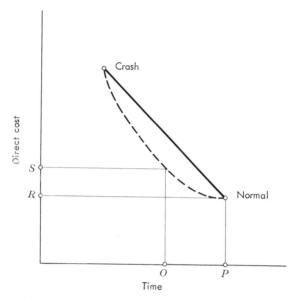

Figure 9–2. Time-cost relationship in which an initial reduction of time can be effected with a modest increase in cost. The approximate curve is the solid line, and the true curve is the broken line.

the increase in cost is slight relative to the decrease in time. You can easily imagine the accounting problems involved in determining the shape of the true time-cost curve (the dotted line), particularly if there are thousands of activities on a project. For this reason, the linear approximation (the straight line connecting the normal and the crash points) is used instead. Although this introduces some error, that error is not really significant in most cases.

Another case of time-cost relationship is illustrated in Figure 9–3. Here the relationship is exactly the reverse of that illustrated in Figure 9–2. A reduction in time from point P to point O is achieved with an increase in cost from point R to point S. The increase in cost is high relative to the decrease in time. We say that this activity is costly to expedite, that its time is expensive to reduce.

If we were trying to reduce the earliest expected date of a project in which the activities pictured in Figures 9–2 and 9–3

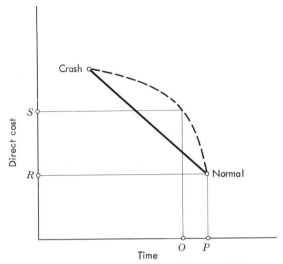

Figure 9–3. Time-cost relationship in which a reduction in time can be effected with a large increase
in cost. The approximate curve is the solid line, and
the true curve is the broken line.

were part of the critical path, we would, of course, first attempt to
crash the activity shown in Figure 9–2 because its time can be
reduced rather inexpensively. On the other hand, reducing the
normal time of the activity illustrated in Figure 9–3 is expensive
and would be attempted only after other activities with more
favorable time-cost curves had been crashed.

The differences between the approximate time-cost curve
(solid line) and the true time-cost curve (dotted line) in Figures
9–2 and 9–3 were intentionally exaggerated. Figure 9–4 illustrates
the more usual case, in which the straight line is a reasonably accurate linear approximation of the true relationship. Two situations,
case *A* and case *B,* are illustrated in Figure 9–4; case *A* illustrates
a time-cost relationship that is *slightly* less favorable than that
illustrated by case *B*. The central idea in using linear approxima-
tions to the true time-cost curves is to be able to determine quickly
the cost of expediting any one of the activities on a network without getting involved in complicated accounting concepts. This is

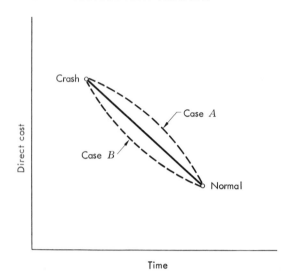

Figure 9–4. The more usual kind of time-cost relationship, in which the solid line is a reasonably accurate linear approximation of the true relationship. Case *A* is slightly less favorable than case *B*.

not to say that we are unable to find the true time-cost curve; we can, but experience with the critical path method has clearly shown that the extra expenditure of time to determine these exact cost relationships is not warranted.

We shall illustrate several activities on an assumed network and give normal and crash figures for each. Then we shall calculate the relative cost of expediting each activity (see Table 9–1). Once again, let us reiterate the purpose of doing all this work: it is to determine where on the critical path substantial reduction in time can be effected for the minimum expenditure of additional money. Given a group of activities lying on the critical path, we can subject them to the analysis illustrated in Table 9–1 with the idea of achieving the greatest reduction in project time with the least increase in project cost. Of course, there is much more to the technique, but this is the central notion.

Expediting the project (often referred to as "crashing" the project) involves choosing from among many alternative methods.

Table 9–1. Calculation of the Relative Costs of Expediting Activities

ACTIVITY	TIME, WEEKS NORMAL	CRASH	COST NORMAL	CRASH	NO. OF WEEKS REDUCTION POSSIBLE	COST INCREASE	COST TO EXPEDITE PER WEEK
1-2	5	3	$10,000	$15,000	2	$ 5,000	$ 2,500
1-3	3	2	15,000	25,000	1	10,000	10,000
3-4	13	9	20,000	24,000	4	4,000	1,000
2-4	11	3	8,000	14,000	8	6,000	750
2-3	6	2	3,000	5,000	4	2,000	500
4-5	3	2	2,000	3,000	1	1,000	1,000
3-5	2	1	6,000	20,000	1	14,000	14,000

$$\text{Cost to expedite per week} = \frac{\text{crash cost} - \text{normal cost}}{\text{normal time} - \text{crash time}}$$

$$\text{or (for activity 2-4)} = \frac{\$14,000 - \$8,000}{11 - 3}$$

$$= \frac{\$6,000}{8}$$

$$= \$750 \text{ per week to expedite}$$

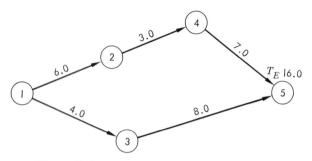

Figure 9–5. Network with all normal times.

Given a set of activities on the critical path of a network, each with its own cost per week of crashing, there is a large number of alternative ways to crash the project, each with its own cost. Let us look now at the network in Figure 9–5 and Table 9–2. Normal times are shown on the network itself. Path 1-2-4-5 is the critical path in this instance, and the earliest expected date of the network-ending event using normal times is 16.0 weeks.

The network that was first introduced in Figure 9–5 is repeated in Figure 9–6 to illustrate crash times for all the activities. From this figure we can see that if we use crash times for all activities the earliest expected date of the network-ending event becomes 10.0 weeks and the critical path is still 1-2-4-5. To crash all activities would result in a total project direct cost of $47,000 (the total of the crash cost column in Table 9–2), but this would deliver the project in 10.0 weeks. At the other extreme, finishing

Table 9–2. Calculation of the Cost of Crashing a Program

	TIME, WEEKS		COST		COST TO REDUCE PER WEEK
ACTIVITY	NORMAL	CRASH	NORMAL	CRASH	
1-2	6.0	4.0	$10,000	$14,000	$2,000
1-3	4.0	3.0	5,000	8,000	3,000
2-4	3.0	2.0	4,000	5,000	1,000
3-5	8.0	6.0	9,000	12,000	1,500
4-5	7.0	4.0	7,000	8,000	333

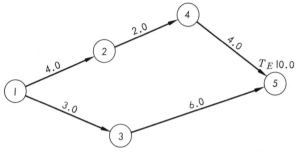

Figure 9–6. Figure 9–5 with crash times added for all the activities.

the project in the normal time would require 16.0 weeks, but the accompanying total direct cost would be only $35,000 (the total of the normal cost column in Table 9–2). A question we should ask ourselves at this point is "Is it possible to reduce the project time to 10.0 weeks without increasing costs $12,000 ($47,000 − $35,000)?" Put another way, the question becomes one of determining whether there is a way of expediting the project without crashing every activity in the network. Fortunately there is a procedure which may accomplish this for us.

Let us begin by simply thinking our way through the problem. We will not use any formalized system or procedure for our simple illustrative network; however, when we have gone through the procedure once we shall then come back to a larger, more complex network and formalize what we learned. This will allow us to cope with networks of larger size and greater complexity.

Referring back to Table 9–2, we notice that the least expensive activity to crash on the critical path is activity 4-5; here time can be reduced for a cost of $333 per week. The minimum crash time to which we can reduce this activity is 4.0 weeks at a total cost of $1,000. When we have accomplished this, the network appears as in Figure 9–7. The earliest expected date of the network-ending event is 13.0 weeks, and total project cost is $36,000.

The next least expensive activity to crash on the critical path is activity 2-4. Reduction of time on this part of the project can be accomplished for a cost of $1,000 per week; the time can be reduced from 3.0 to 2.0 weeks. When this has been accomplished,

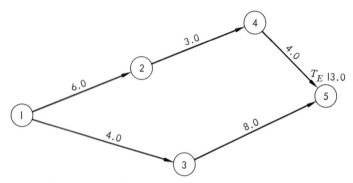

Figure 9–7.   Figure 9–6 with activity 4–5 crashed.

the network appears as in Figure 9–8. The T_E of the network-ending event is 12.0 weeks, and the total project cost is $37,000.

It is apparent from Figure 9–8 that both paths 1-2-4-5 and 1-3-5 are critical; each path requires 12.0 weeks for performance of the work required. Any reduction of the time on one path by further crashing without a corresponding reduction of the time on the other path will not reduce the T_E of the network-ending event any further. For example, suppose that we went ahead and crashed activity 3-5 (the next least expensive) from 8.0 weeks to 6.0 weeks: the lower network path 1-3-5 after crashing requires 10.0 weeks, but the upper path 1-2-4-5 still requires 12.0 weeks; thus the T_E of the network-ending event will still be 12.0 weeks

Figure 9–8.   Figure 9–7 with activity 2–4 crashed.

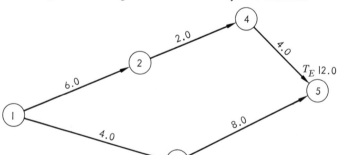

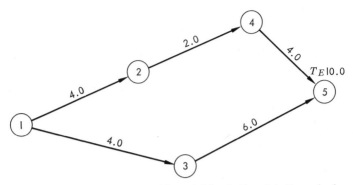

Figure 9–9.   Figure 9–8 with activities 3–5 and 1–2 crashed.

notwithstanding the money we spent to crash activity 3-5. Such action therefore would not be worthwhile.

If, however, we were to crash activity 3-5 from 8.0 weeks to 6.0 weeks and at the same time crash an activity on the upper path the same 2 weeks, then and only then could the T_E of the network-ending event be reduced. Let us proceed in that direction. Crashing activity 3-5 from 8.0 to 6.0 weeks at a total cost of $3,000 ($1,500 per week) and then crashing the only remaining un-crashed activity on the upper path (activity 1-2) from 6.0 weeks to 4.0 weeks at a total cost of $4,000 ($2,000 per week) will result in the network shown in Figure 9–9. In that network, the project can be completed in 10.0 weeks at a total crashed cost of $44,000. This is $3,000 less than the cost of crashing all activities, yet achieves the same results.

At this point one might wonder why we do not go ahead and crash activity 1-3. According to the data in Table 9-2, the time for this activity in the final plan (4.0 weeks) is not the lowest possible crashed time (3.0 weeks). If we did elect to crash activity 1-3 to its lowest possible time, the lower path of the network would total 9.0 weeks and the upper path would still total 10.0 weeks, because all activities on the upper path have been crashed to their lowest time values. To crash activity 1-3 would be to spend $3,000 foolishly, as we would not be able to reduce the T_E of the network-ending event below 10.0 weeks. The recapitulation of what we have done that appears in Table 9–3 may be helpful.

Table 9–3. The Network in Figure 9–5 and Table 9–2 after Crashing

STEPS	T_E OF NETWORK-ENDING EVENT, WEEKS	TOTAL NETWORK COST
1. Original network	16.0	$35,000
2. Crash activity 4-5 to 4.0 weeks	13.0	36,000
3. Crash activity 2-4 to 2.0 weeks	12.0	37,000
4. Crash activity 3-5 to 6.0 weeks	12.0	40,000
5. Crash activity 1-2 to 4.0 weeks	10.0	44,000

Remember that in crashing this project to its lowest possible time at the minimum possible cost, we have taken into account only the *direct* costs associated with the project. Labor, materials, and such are essentially costs that vary directly with the time required to complete the work. Nothing has been said about (1) *indirect* costs (the overhead costs that go on almost irrespective of the time required to complete the work) or (2) costs sometimes referred to as *utility* costs (e.g., penalties for being late and bonuses for finishing the project early). The behavior of these two types of costs can certainly influence the decision about the desirability of crashing a project.

Assume that a contractor in his original contract promised delivery in 12.0 weeks; assume further that the contractor agreed to pay a penalty of $10,000 per week if he delivered later than 12.0 weeks. Thus, when the contractor sees that he will not finish before 16.0 weeks, he faces a possible total penalty of $40,000.

Figure 9–10.   Network used to illustrate crashing procedure.

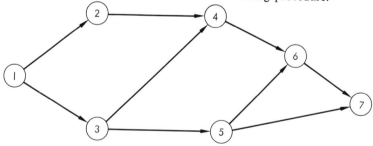

Table 9—4. Costs and Times for Crash and Normal Schedules for the Network in Figure 9–10

ACTIVITY	TIME, WEEKS		COST		COST TO EXPEDITE PER WEEK
	NORMAL	CRASH	NORMAL	CRASH	
1-2	6	2	$ 4,000	$12,000	$2,000
1-3	8	3	3,000	6,000	600
2-4	7	4	2,800	4,000	400
3-4	12	8	9,000	11,000	500
4-6	3	1	10,000	13,000	1,500
5-6	5	2	4,900	7,000	700
3-5	7	3	1,800	5,000	800
5-7	11	5	6,600	12,000	900
6-7	10	6	4,000	8,400	1,100
			$46,100	$78,400	

No doubt he would be glad to incur crashing costs that would reduce the T_E to 12.0 weeks as long as they were less than $40,000. On the other hand, there is no reason why he would want to spend additional money to reduce the project time to under 12.0 weeks unless the reduction in indirect costs is greater than the crashing costs.

There are many possible ways in which these three types of costs, direct, indirect, and utility, can be combined in project planning and control. We shall not at this point illustrate in detail an example of each possible combination. We shall return to this subject for further analysis when we have introduced a more formal method of crashing the larger, more complex type of projects.

To begin our formal method of finding the optimum cost and time duration for a project, we need a network as an example. Such a network has been provided in Figure 9–10. Costs and times for both crash and normal schedules are provided in Table 9–4.

We now set up two networks, one using all normal times for the activities, the other using all crash times. These two complete networks appear in Figures 9–11 and 9–12. In each case the critical path is illustrated with a heavy line.

If the project is completed on a normal basis, it will require

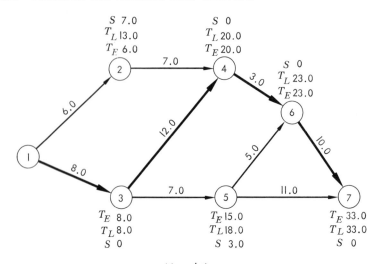

Normal times

Figure 9–11. Figure 9–10, using all normal times for the activities. The critical path is shown by a heavy line.

Figure 9–12. Figure 9–10, using all crash times for the activities. The critical path is shown by a heavy line.

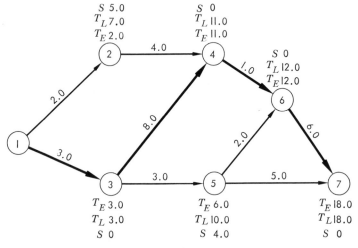

Crash times

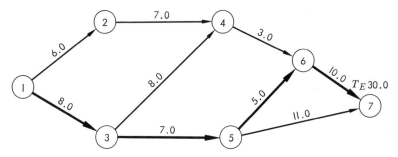

Figure 9–13. New critical path (heavy line) for Figure 9–11, created by crashing activity 3–4.

33.0 weeks and cost $46,100. If all activities are crashed, the project can be completed in 18.0 weeks at a cost of $78,400. Our task is to determine whether we can reduce the time to something near 18.0 weeks without incurring $78,400 in total direct costs. There are several methods for accomplishing this; we have elected to use one which can be broken down into steps and which guarantees that each successive step will be an improvement over the preceding one. Using a logical step-wise procedure makes the procedure easier to program for a computer than if we simply used trial and error.

We begin by observing that the critical path in the all-normal network (Figure 9–11) is path 1-3-4-6-7. The least expensive of the activities on the critical path to crash is activity 3-4 ($500/week). We crash this activity to its minimum time of 8.0 weeks, at a total cost for this action of 4 weeks × $500/week, or $2,000. This creates a new critical path, shown in Figure 9–13 by a heavy line.

The new critical path is path 1-3-5-6-7, with a total expected elapsed time of 30.0 weeks. Proceeding as before, we crash the activity on this new critical path which is least expensive to reduce, activity 1-3 in this case. The time for activity 1-3 is reduced from 8.0 to 3.0 weeks at a total cost for this action of 5 weeks × $600/week, or $3,000. The resulting network is illustrated in Figure 9–14. A new critical path 1-2-4-6-7 has been defined.

Of the activities on the new critical path, activity 2-4 is the least expensive to crash, at $400/week. Proceeding as before, we

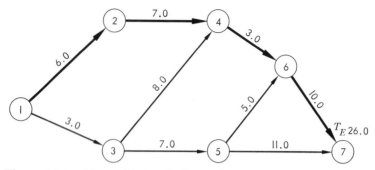

Figure 9–14. New critical path (heavy line) for Figure 9–11, created by crashing activity 1–3.

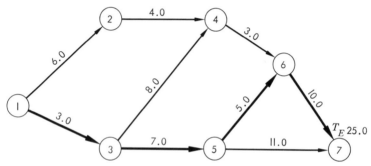

Figure 9–15. New critical path (heavy line) for Figure 9–11, created by crashing activity 2–4.

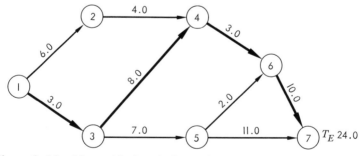

Figure 9–16. New critical path (heavy line) for Figure 9–11, created by crashing activity 5–6.

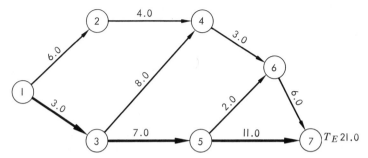

Figure 9–17. New critical path (heavy line) for Figure 9–11, created by crashing activity 6–7.

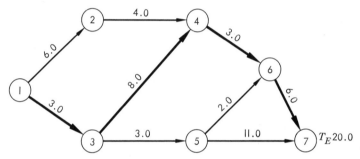

Figure 9–18. New critical path (heavy line) for Figure 9–11, created by crashing activity 3–5.

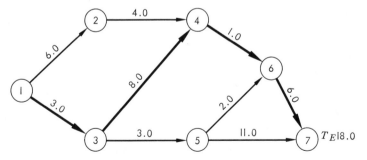

Figure 9–19. New critical path (heavy line) for Figure 9–11, created by crashing activity 4–6.

crash this activity from 7.0 down to 4.0 weeks at a cost of 3.0 weeks × $400/week, or $1,200. This creates a new network and a new critical path, both illustrated in Figure 9–15.

Observing the activities on the new critical path 1-3-5-6-7, we see that activity 5-6 is the least expensive to crash, at $700/week. We crash activity 5-6 to its minimum time of 2.0 weeks at a cost for this action of 3.0 weeks × $700/week or $2,100. The resulting network and critical path are shown in Figure 9–16.

Of the activities on the new critical path 1-3-4-6-7, activity 6-7 is the least expensive to crash at this point. We crash that activity to its minimum time of 6.0 weeks at a cost of 4.0 weeks × $1,100, or $4,400. The new network and critical path are illustrated in Figure 9–17.

Of the activities on the new critical path 1-3-5-7, activity 3-5 is the least expensive to crash; as before, we reduce the time for this activity to its minimum of 3.0 weeks at a cost of 4.0 weeks × $800, or $3,200. The new critical path is illustrated in Figure 9–18.

Observing the activities on the new critical path 1-3-4-6-7, we see that all have been crashed except 4-6. Our choice is thus made for us, and we proceed to crash 4-6 to 1.0 week at a cost of 2.0 weeks × $1,500, or $3,000. The results of this action are shown in Figure 9–19.

None of the activities on the critical path in Figure 9–19 can be reduced further; thus, we have defined the longest nonreducible path through the network. The project cannot be completed in less than 18.0 weeks. Let us stop for a moment at this point and calculate what we have spent by crashing some of the activities. Beginning with the action illustrated in Figure 9–13, Table 9–5 shows the results of our crashing actions.

We have thus been successful in reducing the time of the project to 18.0 weeks without increasing the cost to the all-crash figure of $78,400. But perhaps more analysis may allow us to reduce the project cost further without increasing the time beyond 18.0 weeks.

Observe the network in Figure 9–19, the final network that resulted from our crashing activities. Since only four activities are on the critical path, perhaps the activities not on the critical path (and there are five of these) can be "uncrashed" (their time in-

Table 9–5. Results of the Crashing Actions
in Figures 9–13 to 9–19

Original all-normal project cost	$46,100
Crash activity 3-4 to 8.0 weeks	2,000
Crash activity 1-3 to 3.0 weeks	3,000
Crash activity 2-4 to 4.0 weeks	1,200
Crash activity 5-6 to 2.0 weeks	2,100
Crash activity 6-7 to 6.0 weeks	4,400
Crash activity 3-5 to 3.0 weeks	3,200
Crash activity 4-6 to 1.0 week	3,000
Total cost of crashed network	$65,000

creased) with some further saving. Let us attempt to do this beginning logically with the activity that cost the most to crash (activity 1-2 at $2,000/week) and thus the one that will save the most if we are successful in uncrashing it. From this beginning point we should proceed to the next most expensive activity and so on until we are certain that no further uncrashing is possible. Then and only then can we say that we have achieved the min-

Table 9–6. Results of Uncrashing Activities Not on the Critical Path

NONCRITICAL PATH ACTIVITY	VALUE PER WEEK IF UNCRASHED	ACTION TAKEN	NET GAIN REALIZED
1-2	$2,000	None; cannot be uncrashed further	$ 0
5-7	900	None; cannot be uncrashed further	0
3-5	800	Time extended to 4.0 weeks; path 1-3-5-7 will not permit further extension	800
5-6	700	Time extended to 5.0 weeks; 5.0 weeks is the normal time for this activity	2,100
2-4	400	Time extended to 5.0 weeks; path 1-2-4-6-7 will not permit further extension	400
Total gain from uncrashing noncritical path activities			$3,300

imum project cost commensurate with the time of 18.0 weeks. This procedure is demonstrated in Table 9–6. At this point, further reduction of project cost is impossible if we are to finish in 18.0 weeks. Our analysis is complete, and the resulting network is pictured in Figure 9–20. The cost of the project at this point is the $65,000 crashed network cost minus the uncrashing savings of $3,300, or $61,700. Notice that all paths through the network in Figure 9–20 are critical (they all have total expected elapsed times of 18.0 weeks).

Let us review briefly the procedure for determining the minimum direct cost. We begin by observing the all-normal network to determine the critical path and the least expensive activity on the critical path to crash. We then crash the least expensive activity to its crash time. This defines a new critical path with its own least expensive activity to crash. We proceed in this manner until a critical path that is not reducible (a path on which all activities are at their crash times) has been defined. We then reverse the procedure by observing the activities not on the critical path and uncrashing them, beginning with the most expensive activity to crash and proceeding to the least expensive activity. We uncrash each of these as far as the critical path will permit us; i.e., to the point where further uncrashing would create a new (longer) critical path, or to their normal times, whichever is shorter. This completes the procedure.

One can appreciate how involved this procedure would become for a larger network. Many steps and calculations would be

Figure 9–20. Figure 9–19 after all activities not on the critical path have been uncrashed.

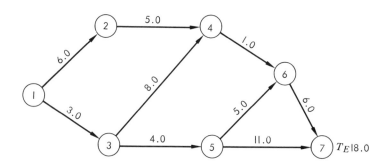

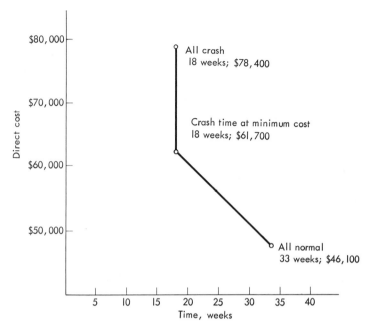

Figure 9–21. Total direct cost plotted against total time necessary to complete the work.

involved, and the procedure could become quite tiresome. Most CPM networks involving more than a very few events are solved (crashed and uncrashed) with the help of a computer. Using somewhat the same procedure we have outlined, the computer can yield the correct results in a matter of minutes whereas hand solution would require days.

We can illustrate the cost import of what we have done by plotting the direct cost curve of the project. We simply plot the total direct cost against the total time (in weeks) necessary to complete the work. This concept is illustrated in Figure 9–21. The curve in Figure 9–21 shows that spending more than $61,700 is foolish because the time is not reducible below 18.0 weeks. The reverse holds true also: Taking longer than 33.0 weeks to do the work is equally foolish because cost does not drop below $46,100.

But what about the other costs we have mentioned briefly:

indirect costs and utility costs? We saw that they can possibly affect the extent (if any) to which it is profitable to crash a project. Indirect costs are such expenditures as rental on equipment, fixed costs of supervisory salaries, costs of payment and performance bonds on projects, and the like. Utility costs are usually limited to bonuses or penalties for finishing either early or late. We can illustrate from Figure 9–21 how these three costs, direct, indirect, and utility, interact to determine the optimum time duration of a project. We assign to this project $1,000 per week as an indirect cost and $1,500 per week as a penalty cost for completing the project in more time than 20 weeks. All three costs are plotted on the graph in Figure 9–22. In addition, the project total cost curve has been constructed by adding together the three previously mentioned curves.

You will notice that the total project cost curve is lowest at a project completion time of about 20.0 weeks. This means that because of the interaction of *all* the costs, it is more advantageous for the company to allow the project to require 20.0 weeks than to finish it in 18.0 weeks. The logic behind this reasoning is not difficult to follow. To reduce the project from 20.0 weeks to 18.0 weeks costs. When the project manager tried to get the last 2 weeks' possible reduction, he would, of course, be dealing with the most expensive activity to expedite, i.e., activity 4-6 at $1,500 per week. Why? Because he would have exhausted long ago all the activities that are less expensive to expedite. Now what has he to gain by expediting from 20.0 weeks to 18.0? Only a savings of $1,000 indirect costs per week because the penalty does not even start until the 20th week. No wonder our total project cost curve tells us that about 20.0 weeks is the optimum schedule. It is not optimum from the point of view of direct costs alone, but it is optimum when all three types of costs associated with the project are taken into consideration.

How should our project manager arrange the project so that it is completed in exactly 20.0 weeks? Surely there are many possible combinations of activity times between all crash and all normal that would generate a total project time of 20.0 weeks. How can we pick the least costly method of finishing in this time?

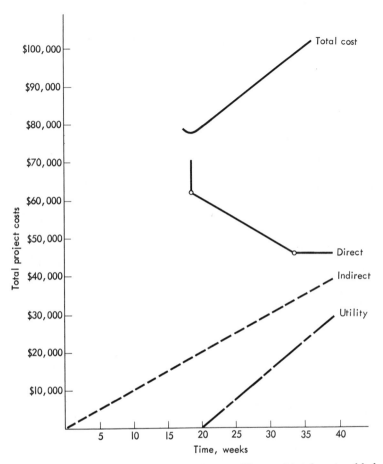

Figure 9–22. Figure 9–21 with indirect, utility, and total costs added.

Let us go back for a moment to activity 4-6, the last one we crashed when we were attempting to reduce the network project time. This was reserved for last since it was the *most* expensive activity to crash. If you were the project manager then and were told you could take 2 weeks longer, would you not lengthen (uncrash) the activity that cost the most to crash? Thus, the optimum response will be to let activity 4-6 be completed in 3.0

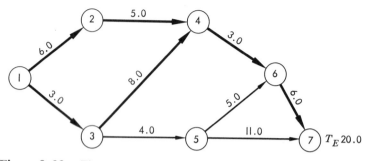

Figure 9–23. Figure 9–20 after activity 4–6 has been uncrashed to complete the network in optimum time.

weeks instead of 1.0 week. The resulting network is illustrated in Figure 9–23. Notice that there are two critical paths in this network, paths 1-2-4-6-7 and 1-3-4-6-7, and three activities not on the critical path. With a new critical path (or with two new critical paths, in this case) defined, we might attempt to uncrash further the activities that are not on the critical path to save additional expense. Of the three activities (3-5, 5-6, and 5-7) not on the critical path, two (5-6 and 5-7) have already been uncrashed to the normal time. This leaves only activity 3-5 which could be uncrashed. The project manager would of course uncrash activity 3-5 to 6.0 weeks, thereby saving an additional 2 weeks × $800/week, or $1,600. The final network appears in Figure 9–24.

The final project cost, $77,100, is calculated as follows: $61,700 original direct cost, less $3,000 recovered from uncrash-

Figure 9–24. Figure 9–23 after activity 3–5 has been uncrashed to complete the network in optimum time.

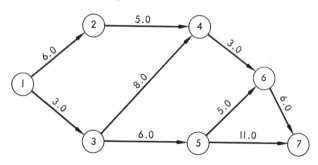

ing activity 4-6, less $1,600 recovered from uncrashing activity 3-5, plus $20,000 indirect costs.

All the work we have done to arrive at an optimum total project cost should be done as forward planning before a project is begun. It must be done again each time some change that significantly alters the completion date occurs while the work is going on. Many construction projects replan the network on a monthly basis. With a computer this is not nearly the task it would be with hand computations.

Other Methods
of Project Planning
and Control

Methods of planning and control other than PERT have been developed over the years. Users of PERT have found that modifications of and newer developments in the original PERT technique have generated better results or have enabled project managers to achieve greater efficiency.

Many new systems have been designed since PERT was originally used on the fleet ballistic missile program. An estimate of the total number of revised systems might go as high as 100. True, they all in one sense or another have their foundation in the simple ideas of networking associated with the original PERT system, but all have added some features not found in the original PERT idea.

A few of the important systems in current use are discussed below, systems that are, in one manner or another, refinements of PERT. Our treatment of each in this small book must be extremely short, even though an entire chapter could easily be devoted to each. Despite the brevity of the descriptions, we think that our references to other methods warrant inclusion. There is value in knowing about other approaches and the directions they are taking.

1.   MANPOWER UTILIZATION

Manpower Utilization, a fairly recent development in the area of project planning and control, concerns the optimum assignment, utilization, and control of manpower within projects. Using this technique, management can assign available manpower within projects and predict the probable effects upon project objectives if workloads and assignments are altered. This special modification of PERT allows the user to concentrate upon scientific manpower as the single most important resource. Considering the tremendous number of technical and scientific persons required on some of our defense projects, we can surely see that individuals *could* easily represent a major constraint and that a specialized technique of manpower utilization can be of tremendous value to a project manager.

2.   LINE OF BALANCE

Line of Balance (commonly referred to as LOB) is a useful addition to PERT. It is a system for planning and controlling production that concentrates on the time aspects of *key* events required for completion of the entire project; it does not ordinarily go into the detail present in PERT. Graphic displays are used to monitor the progress achieved on the project and to illustrate where some objective is not being realized. In this manner the graphic display allows a certain amount of management by exception and thus is quite useful.

Frequently PERT is used to plan and control the project up to the point where the first article or unit is completed. Thus, PERT is used during the research and development stage, in which the greatest uncertainty concerning time values exists. When substantial production of the article begins, Line of Balance may represent a more appropriate alternative for future project control for several reasons. Line of Balance can generate historical information faster, which in turn can be used in further reduction of project lead times. In general, LOB is not so involved or so detailed as PERT; it does not require the use of a computer, and thus may be considerably less expensive.

3. IMPLEMENTATION, PLANNING AND CONTROL TECHNIQUE

The Implementation, Planning, and Control Technique (IMPACT) is a networking system designed specifically to help management control the cost of preparing and installing computer programs. Many executives have found that installation of the "hardware" (the computer itself and allied equipment) was the easiest part of getting into the space age computationally. The planning and control of the "software" (i.e., the writing of programs that will make the machine operational) is the real time-consuming activity in the process.

IMPACT was originally designed to tell management what the programming would probably cost, how long it would take, and the manpower requirements it would consume. IMPACT utilizes a network on which the events represent specific accomplishments during the programming process. Time is measured in hours.

4. CONTRACTUAL REQUIREMENTS RECORDING, ANALYSIS, AND MANAGEMENT

The Contractual Requirements Recording, Analysis, and Management system (CRAM) was specifically designed for the purchasing operation in Air Force projects. When applied to the purchasing area usually called "procurement follow-up" (i.e., monitoring the procurement action from the time the purchase order is issued until the purchased article is received into stock), CRAM is said to reduce the work involved and to have a desirable effect upon the time required for the procurement cycle.

Each activity in the purchasing process is allotted a standard amount of time, and each type of contract is allowed a standard amount of time for processing. Given the technical details of each anticipated procurement action, the computer develops a series of cards representing each phase (activity) in the procurement action. The appropriate card is sent to the person responsible for each of these activities, and is returned to the computer center when the action indicated has been completed. Periodically, perhaps each week, the computer takes all cards that have come in and estimates the completion dates of all the procurement activities that follow, then forecasts the completion date of the contract. It computes the time position of the contract (number of days ahead of schedule, number of days behind schedule, what has been completed, even reasons for delay) and prepares a printed report for each contract. Thus, each buyer in the procurement office involved gets a periodic report on the contracts for which he is responsible, and the management of the procurement operation is able to monitor progress on each of the contracts involved without getting involved in every detail of the process. CRAM is particularly helpful on projects in which thousands of individual contracts have to be negotiated with suppliers and follow-up action maintained to ensure timely completion of the overall project. Without some automated system for showing the current status of each of the contracts and the effects of each of them upon the project completion date, management of such an involved project would indeed be complicated.

5. LEAST COST ESTIMATING AND SCHEDULING

The Least Cost Estimating and Scheduling technique (often called LESS) is a networking procedure that determines the fastest *and* the most economical method of completing a project. Obviously, when we speak of economics we realize that LESS must assign not only a time value to each of the activities but also a cost; this is quite a departure from the original PERT idea. This idea was originally developed by International Business Machines Company. Because LESS was designed for the smaller computers, the scope of the projects on which it can be utilized is somewhat limited. The critical factor in LESS is the introduction of cost; as we saw in Chapter 9, the introduction of cost to the PERT process represents a significant development in networking techniques and makes them of even greater value to management in the planning and control of projects.

6. COMPUTER OPERATED MANAGEMENT EVALUATION TECHNIQUES

Computer Operated Management Evaluation Techniques (COMET, as it is commonly known) is a variation of PERT. Unlike the Air Force, which might have only a few missile systems in the process of development at one time, the Army Material Command has hundreds of items of equipment in development simultaneously; it must therefore exercise a reasonable degree of control over all these projects. The Army Material Command has effectively substituted COMET and other computer-based systems for control by staff officers.

An interesting feature of COMET is its use of a standard network, one which will suffice for any type of equipment procurement project. The standard network is of a size that will illustrate even the most complex developmental program; yet it can be adapted for much smaller programs by the use of the zero-time activity concept developed in Chapter 5.

7 TO 30

Each of the following modifications is defined briefly. The brief descriptions are given only for reference purposes. Persons wishing to learn more about the techniques mentioned can, of course, go directly to the sources mentioned.

7. ACTIVITY BALANCE LINE EVALUATION

The Activity Balance Line Evaluation (ABLE) is a program-status measuring, forecasting, and reporting system. The name is derived from the summary of management presentation where accumulative summaries of processes are shown in balance against a "today" line on the time field. Source: Missile and Space Division, General Electric Company, Philadelphia, Pa.

8. BUWEPS PERT MILESTONE SYSTEM

By combining advanced management techniques with the full utilization of the computer capability, the BUWEPS PERT milestone system is designed to improve communications for facilitating the decision-making process in the Navy-contractor team endeavor for developing new weapon systems. Source: U.S. Navy Department, Progams Office, Washington, D.C.

9. COST PLANNING AND APPRAISAL

Cost Planning and Appraisal (CPA) is designed to assist equipment managers in developing a more effective means of managing cost-type contracts by integrating data on cost, time, and technology. Source: Aeronautical Systems Division, Air Force Systems Command, Wright-Patterson Air Force Base, Ohio.

10. HOFFMAN EVALUATION PROGRAM AND PROCEDURE

The Hoffman Evaluation Program and Procedure (HEPP) is an activity-oriented planning and control system tailored especially for research and development work. It presents the following prime elements of a program plan on a single graphic display: work activities, hardware, time allocation of resources, and costs. Frequent updating shows what has happened since the last review period and what departures from the plan are anticipated. Source: Military Products Division, Hoffman Electronics Corporation, Los Angeles, Calif.

11. INTEGRATED CONTROL

Integrated Control (ICON) is a management information system which is used (1) in the preparation and dollar evaluation of bids and in periodic reevaluation after contract award, (2) for scheduling through PERT techniques including a network reduction algorithm and complete master file maintenance, and (3) for complete data collection and reporting of actual versus planned expenditure of time, dollars, and manpower. Source: Sylvania Electronics Systems Division, Sylvania Electric Products, Inc., Needham Heights, Mass.

12. MANAGEMENT PLANNING AND CONTROL SYSTEM

The Management Planning and Control System (MPACS) is the basic communication of financial and manpower data—budget versus actual—from the accounting or data-collecting agency to the program manager who is responsible for contract performance and the performing functional manager. It provides data weekly, three working days after the close of the previous week. With the advent of PERT/Cost, MPACS was reconstructed to include PERT/Cost terminology. It thus serves as the bridge between the detail cost within a Work Package and PERT/Cost's summary figure of "actual costs." It also provides the work package manager with the inception-to-date expended and committed data so that he can better prepare the PERT/Cost Estimating and Updating Form. It uses 1410, 1401, and 704 computers on all government contracts. Source: Federal Systems Division, International Business Machines Corporation, Rockville, Md.

13. NASA PERT AND COMPANION COST

The NASA PERT and Companion Cost system is designed as a total management system concept. It integrates the normal existing NASA management and administrative tools and processes into a disciplined planning, control, and reporting instrument for the project manager. The basic theme is that total project management can be achieved only if the three management variables—time, resources, and performance—are managed and manipulated on a common framework which classifies all work elements of the project beginning from the top and spreading out to the successive tiers representing systems, subsystems, etc., which make up the project. Source: Director of Management Reports, Office of Programs, NASA, Washington, D.C.

14. PROGRAM ANALYSIS ADAPTABLE CONTROL TECHNIQUE

The Program Analysis Adaptable Control Technique (PAAC) is a method of extracting, analyzing, and summarizing planned and actual data contained in the engineering management information system independent of the existing work breakdown structures used for individual project control purposes. It makes possible anticipation of the control needs of both customer and program manager, integration of these into the overall reporting system, and then automatic recovery of the data to support these needs. It uses a Honeywell-800 computer as its central data-processing instrument. Source: Military Products Group Aeronautical Division, Minneapolis-Honeywell Regulator Company, Minneapolis, Minn.

15. PERT/COST

PERT/Cost is an extension of the basic PERT/Time system for the management of complex research and development projects to achieve their technical program objectives. Both cost and schedule are planned and controlled on a common framework or structure, which not only permits more accurate measurement of progress but also enables managers to appraise more realistically the relation of accumulated and projected costs of the program. In addition, it provides insight for alternative courses of action as time and cost are affected. PERT/Cost is considered to be in developmental status since it is being tested by the Army, Navy, Air Force, and other organizations. Source: Air Force Systems Command (SCCS), Andrews Air Force Base, Washington, D.C.

16. PROGRAM EVALUATION PROCEDURE

The Program Evalution Procedure (PEP) is the original Air Force version of PERT. A management information and control system used in the planning, control, and evaluation of progress of a program, it is time-oriented and uses a network to reflect events and activities. The basic system was designed by the Navy for its Polaris program. It is now being used by the Department of Defense, NASA, and other government agencies and industries. Source: Air Force Systems Command (SCCS), Andrews Air Force Base, Washington, D.C.

17. PERT II

PERT II is a version of PERT used by the Ballistic Systems Division, Air Force Systems Command, in the Minuteman program. Source: Air Force Systems Command (SCCS), Andrews Air Force Base, Washington, D.C.

18. PERT III

PERT III is the Air Force Systems Command's Standard PERT/ Time Variable System. It integrates PERT (U.S. Navy), as amended and applied by the Aeronautical Systems Division and the Electronic Systems Division, and PERT II (Minuteman), as applied by the Ballistic Systems Division and the Space Systems Division, for uniform application of the PERT/Time Variable System within the command. Source: Air Force Systems Command (SCCS), Andrews Air Force Base, Washington, D.C.

19. PERT IV

PERT IV is the Air Force Systems Command's Standard PERT/ Time and Cost Variables System. It integrates PERT III and enumerative cost aspects for uniform application of the PERT/Time and PERT/Cost systems within the command. Source: Air Force Systems Command (SCCS), Andrews Air Force Base, Washington, D.C.

20. PROJECT AUDIT REPORT

Project Audit Report (PAR) is a modified PERT/Cost system, based upon the presently used systems within the Burroughs Detroit Division, used to fulfill the need for increased initial planning as well as prompt and accurate reporting techniques. It provides an extremely effective tool for time/cost reporting of projects of any magnitude, affording a documented, logically ordered history of past performance, which, when statistically evaluated and analyzed, will serve in sharpening abilities for future project planning and cost estimating. Source: Detroit M&E Division, Burroughs Corporation, Detroit, Mich.

21. PLANNING NETWORK

Planning Network (PLANNET) is a scheduling technique for planners which is used in the Guided Missile Range Division of Pan American World Airways. It is visualized scheduling combining bar charts, in series and in parallel, on a time-oriented chart. It is similar and supplementary to PERT. Source: Guided Missile Range Division, Pan American World Airways, Patrick Air Force Base, Fla.

22. PROGRAM RELIABILITY INFORMATION SYSTEM FOR MANAGEMENT

The Program Reliability Information System for Management (PRISM) is a system for developing new management tools for monitoring and measuring reliability for the Navy Polaris program, as well as for improving communications. It uses two unrelated approaches: (1) RMI—Reliability Maturity Index, a measure of compliance with the planned reliability activities of the development program, and (2) RPM—Reliability Performance Measure, a prediction of eventual operational reliability of the end item continuously through the development cycle. The methodology employed is an offshoot of the PERT system. Source: U.S. Navy Department, Washington, D.C.

23. RESOURCES ALLOCATION AND MULTI-PROJECT SCHEDULING

Resources Allocation and Multi-Project Scheduling (RAMPS) is an automated management technique for making the most of men, materials, and money. It is based on CPM and PERT, and was developed jointly by CEIR, Inc., and E. I. Du Pont de Nemours & Company. Source: CEIR, Inc., Arlington, Va.

24. SCHEDULING AND CONTROL BY AUTOMATED NETWORK SYSTEMS

Scheduling and Control by Automated Network Systems (SCANS), unlike basic PERT, uses four criteria: time, labor, facilities, and cost.

SCANS uses these criteria in planning future activities, analyzing present success, and studying the effects of change. SCANS permits a more detailed study than PERT when required; its major advantage over other networking systems is that it brings resource allocation into the determination. Source: *IRE Transactions on Engineering Management,* vol. EM-9, no. 3, September, 1962.

25. COMPUTER PROGRAM FOR SCHEDULING TIME AND DISTRIBUTING COST

Computer Program for Scheduling Time and Distributing Cost (SKED) is an integrated computer program for scheduling time and distributing cost. It is a standard PERT operation, with the addition of a scheduling option based on certain decision rules concerning the allocation of negative and positive slack. Once the slack is allocated, the man-hours and dollars are estimated and the distribution of both is evaluated by management. This system is in the development stage. Source: Finance Section, Ordnance Department, General Electric Company, Pittsfield, Mass.

26. SCHEDULE PERFORMANCE EVALUATION AND REVIEW TECHNIQUE

The Schedule Performance Evaluation and Review Technique (SPERT) is a program control technique used by the Missile and Space Vehicle Department of General Electric. It is based on PERT, but all events in the network are scheduled with firm dates from the commitment plan, and trend projection and management display methods are utilized from the ABLE system. Source: Missile and Space Division, General Electric Company, Philadelphia, Pa.

27. TRADE-OFF EVALUATION SYSTEM

The Trade-Off Evaluation System (TOES) is an advanced management tool being developed by AVCO to permit an efficient rational evaluation of technical characteristics, schedules, and cost trade-offs. It is aimed at establishing compatibility between PERT and systems analysis, recognizing the conflicting nature of the TCP constraints. Source: Research and Advanced Development Division, AVCO, Washington, D.C.

28. THE OPERATIONAL PERT SYSTEM

The Operational PERT System (TOPS) is a generalized PERT system showing only one network. It was developed by Aerospace Corporation for use on Air Force Space System Division programs. Source: Aerospace Corporation, El Segundo, Calif.

29. TASK REPORTING AND CURRENT EVALUATION

Task Reporting and Current Evaluation (TRACE) is a reporting system developed by Ling-Temco-Vought, Inc., for the timely and accurate status reporting of complex programs and the successful management of large and widely dispersed activities. It uses a network of leased lines, data transceivers, and a centralized computing facility. Source: Range Systems Division, Chance Vought Corporation, Dallas, Texas.

30. WEAPON SYSTEMS PROGRAMMING AND CONTROL SYSTEM

The Weapon Systems Programming and Control System (WSPACS) is a computerized force structure costing technique for broad planning with major emphasis on assessing the impacts of reprogramming actions. It is currently organized as a joint Air Force–industry effort. Source: Aeronautical Systems Division, Air Force Systems Command, Wright-Patterson Air Force Base, Ohio.

Bibliography of Books, Pamphlets, and Articles

"Appraisal of Program Evaluation Review Technique, An," *Journal of Academy of Management,* April, 1962.

Archibald, R. D., "PERT and the Role of the Computer," *Computers and Automation,* vol. 12, pp. 26–30, July, 1963.

Astrachan, A., "Better Plans from the Study of Anatomy of an Engineering Job," *Business Week,* Mar. 21, 1959.

"Automated PERT," *Bulletin of the Operations Research Society of America,* Spring, 1962.

Avots, Ivars, "The Management Side of PERT," *California Management Review,* Winter, 1962.

Baker, Bruce N., and R. L. Eris, *An Introduction to PERT/CPM,* Richard D. Irwin, Inc., Homewood, Ill., 1964.

Battersby, Albert, *Network Analysis for Planning and Scheduling,* St. Martin's Press, Inc., New York, 1964.

Beller, W., "Contractor Gets Steady Check on Profits with New PERT-type System," *Missiles and Rockets,* vol. 14, Jan. 13, 1964.

Blickstein, S., "How to Put PERT into Marketing (and Aid Planning)," *Printer's Ink,* vol. 289, pp. 27–29, Oct. 23, 1964.

Bock, Robert H., and W. K. Holstein, *Production Planning and Control*, Charles E. Merrill Books, Inc., Columbus, Ohio, 1963.

Boehm, G. A. W., "Helping the Executive to Make Up His Mind (Decision Theorists Come to Rescue with CPM and PERT Linear Programming, Etc.)," *Fortune,* vol. 65, pp. 128–131+, April, 1962.

Borklund, William, "Why Polaris Is Winning Its Race against Time," *Armed Forces Management,* December, 1958.

Boulanger, David G., "Program Evaluation and Review Technique," *Advanced Management,* July-August, 1961.

Brown, Edward A., *Generation of All Shortest Paths of a Directed Network,* IBM Research Center, Yorktown Heights, N.Y.

Burgher, P. H., "PERT and the Auditor (Program Evaluation and Review Technique)," *Accounting Review,* vol. 39, pp. 103–120, January, 1964.

Carter, J. H., Jr., and G. E. Peek, "Subcontract Control Program (Use of PERT with Companion Cost System)," *NAA Bulletin,* vol. 46, pp. 41–44, July, 1965.

Case, J. G., "PERT: A Dynamic Approach to Systems Analysis," *NAA Bulletin,* vol. 44, pp. 27–38, March, 1963.

Caleo, R. L., "PERT and You," *Administrative Management,* vol. 23; pp. 13–15, March, 1962.

Clark, C. E., "The Optimum Allocation of Resources among the Activities of a Network," *Journal of Industrial Engineering,* January-February, 1961.

Clingen, C. T., "Modification of Fulkerson's PERT Algorithm," *Operations Research,* vol. 12, pp. 629–632, July, 1964.

Collins, F. Thomas, *Practical Applications and Examples of CPM and PERT,* Know-How Publications, Eugene, Ore., 1964.

"Commercial Uses of PERT Increasing, Survey Shows; with Chart," *Steel,* vol. 154, p. 31, Apr. 6, 1964.

Connell, W. S., and W. A. Dreher, "Rate Book Planning (with Aid of PERT)", *Best's Insurance News (Life Edition),* vol. 64, pp. 20–21+, June, 1963.

"Considerations of PERT Assumptions," *Bulletin of the Operations Research Society of America,* Spring, 1962.

"A Cost Control Extension of the PERT System," *Bulletin of the Operations Research Society of America,* Spring, 1962.

"Critical-path Scheduling," *Chemical Engineering,* Apr. 16, 1962.

"Critical-path Scheduling," *Plant Administration and Engineering,* October, 1961.

Davis, G. B., "Network Techniques and Accounting; with an Illustration," *NAA Bulletin,* vol. 44, pp. 11–18, May, 1963.

"Does PERT Work for Small Projects?" *Data Processing,* December, 1962.

Donaldson, W. A., "Estimation of the Mean and Variance of a PERT Activity Time," *Operations Research,* vol. 13, pp. 325–328, Discussion, H. Coon, pp. 386–387, May, 1965.

Dooley, A. R., "Interpretations of PERT (Useful Classification of Selected Readings for Managers)," *Harvard Business Review,* vol. 42, pp. 160–162+, March, 1964.

Dooley, Arch R., and others, *Operations Planning and Control,* John Wiley & Sons, Inc., New York, 1964.

Evarts, Harry F., *Introduction to PERT,* Allyn and Bacon, Inc., Boston, 1964.

Extensions and Applications of PERT as a System Management Tool, Operations Research, Inc., Los Angeles, March, 1961.

Fazar, W., "Origin of PERT," *Controller,* vol. 30, pp. 598–600+, December, 1962.

Fazar, Willard, *PERT Time vs. Resources vs. Technical Performance,* abstract of presentation, IBM Corporation Educational Center, Poughkeepsie, N.Y., Mar. 15, 1961.

Fazar, Willard, "Planning Implementation and Appraisal through 'PERT'," *Business Budgeting,* January, 1962.

Fazar, Willard, "Practical Considerations in Management's Use of PERT," presentation to Inter-agency Task Force on R&D Progress Reporting, Washington, D.C., April, 1960.

Fazar, Willard, "Program Evaluation and Review Technique," *American Statistician,* April, 1959.

Federal Electric Corporation, *A Programmed Introduction to PERT,* John Wiley & Sons, Inc., New York, 1963.

Frambes, Roland, "Next Big Step for PERT," *Aerospace Management,* October, 1961.

Francis, Harold G., and J. Pearlman, "PERT: Program Evaluation and Review Techniques," *Functional Information Bulletin,* Operations Research and Synthesis Consulting Service, General Electric Company, New York.

Freeman, Raoul J., "A Generalized PERT," *Operations Research,* March-April, 1960.

Fulkerson, D. R., "A Network Flow Computation for Project Cost Curves," *Management Science,* January, 1961.

Fulkerson, D. R., "Expected Critical Path Lengths in PERT Networks," *Operations Research,* vol. 10, pp. 808–817, November, 1962.

Fundamentals of Network Planning and Analysis, Booklet no. PX-1842A, Management Systems Department, Remington Rand UNIVAC Division, Sperry Rand Corporation, St. Paul, Minn., 1961.

Gisser, P., Taking the Chances Out of Product Introductions (Using the PERT Technique)," *Industrial Marketing,* vol. 50, pp. 86–91, May, 1965.

Glassford, W. B., "Critical Path Scheduling," *Plant Administration and Engineering,* October, 1961.

Gorham, W., *An Application of a Network Flow Model to Personnel Planning,* USAF Project Research Memorandum RM-2587, The RAND Corporation, Santa Monica, Calif., 1960.

Groelinger, Herbert J., *An Exercise in PERT Planning/Scheduling/Analysis,* Operations Research, Inc., Santa Monica, Calif., 1961.

Gross, Herbert L., "Program Evaluation and Reporting Technique," *Datamation,* February, 1962.

Hamlin, Fred, "How PERT Predicts for the Navy," *Armed Forces Management,* July, 1959.

Hartung, L. P., and J. E. Morgan, "PERT/PEP . . . A Dynamic Project Control Method," FSD Space Guidance Center, IBM Corporation, Owego, N.Y., January, 1961.

Healy, Thomas L., "Activity Subdivision and PERT Probability Statements," *Operations Research,* May-June, 1961.

Heinzel, C., "PERT Technique Can Aid in Annual Report Preparation," *Public Relations Journal,* vol. 20, pp. 11–14, April, 1964.

Houk, L. E., "Cost Side of NASA PERT and Companion Cost," *NAA Bulletin,* vol. 46, pp. 33–40, July, 1965.

Introduction to PERT, An, Headquarters, Air Research and Development Command, USAF, Andrews Air Force Base, Washington, D.C., May 16, 1960.

Jodka, J., "PERT: A Recent Control Concept," *NAA Bulletin,* vol. 43, pp. 81–86, January, 1962.

Jodka, John, "PERT (Program Evaluation and Review Technique): A Control Concept Using Computers," *Computers and Automation,* March, 1962.

Kelley, J. E., Jr., and M. R. Walker, *Critical-path Planning and Scheduling: An Introduction,* Mauchley Associates, Ambler, Pa., 1959.

Kelley, James E., Jr., and Morgan R. Kelley, "Critical-path Planning and Scheduling," 1959 Proceedings of the Eastern Joint Computer Conference.

Kelley, James E., Jr., "Critical-path Planning and Scheduling: Mathematical Basis," *Operations Research,* May-June, 1961.

Klass, P. J., "NASA, Department of Defense Adopt Standard PERT Form (PERT/Cost System)," *Aviation Week & Space Technology,* vol. 76, p. 31, June 11, 1962.

Klass, Philip J., "PERT/PEP Management Tool Use Grows," *Aviation Week,* Nov. 28, 1960.

Klein, H. E., "Psychoanalysis on the Production Line (PERT System)," *Dun's Review and Modern Industry,* vol. 79, pp. 54–55+, February, 1962.

LaFond, L. D., "Fairplan Aids Fairchild Comeback; PERT Techniques," *Missiles and Rockets,* vol. 11, pp. 28–30, Aug. 13, 1962.

Levy, F. K., and others, "ABC's of the Critical Path Method; Excerpts from Industrial Scheduling," *Harvard Business Review,* vol. 41, pp. 98–108, September, 1963.

Lewis, James, "Where PERT Is Headed," *Armed Forces Management,* July, 1961.

Lockyer, K. G., *An Introduction to Critical Path Analysis,* Pitman Publishing Corporation, New York, 1964.

McNeil, J. F., "Program Cost Control Systems (Use of PERT and Supplemental Techniques.)," *NAA Bulletin,* vol. 45, pp. 11–20, January, 1964.

MacCrimmon, K. R., and C. A. Ryavec, "Analytical Study of the PERT Assumptions," *Operations Research,* vol. 12, pp. 16–37, January, 1964.

Malcolm, D. G., "PERT: A Designed Management Information System," *Industrial Management,* June, 1961.

Management Implications of PERT, The, Booz, Allen & Hamilton, New York, 1963.

"Management Side of PERT, The," *California Management Review,* Winter, 1962.

Martino, R. L., "Concepts of PERT/CPM as Part of a Dynamic System of Project Planning, Scheduling, and Control," *Industrial Development and Manufacturers Record*, vol. 134, pp. 97–98, January, 1965.

Martino, R. L., "How 'Critical-path' Scheduling Works," *Canadian Chemical Processing*, February, 1960.

Martino, R. L., "New Way to Analyze and Plan Operations and Projects Will Save You Time and Cash," *Oil/Gas World*, September, 1959.

Martino, R. L., "Problems with PERT (Arrow-diagraming Concept)," *Chemical Week*, vol. 95, pp. 101, July 18, 1964.

Martino, R. L., "What's the Shortest Path in Project Planning?" *Executive Magazine*, August, 1960.

Miller, Robert W., "How to Plan and Control with PERT," *Harvard Business Review*, vol. 40, pp. 93–104, March-April, 1962.

Miller, Robert W., *Schedule, Cost, and Profit Control with PERT: A Comprehensive Guide for Program Management*, McGraw-Hill Book Company, New York, 1963.

Moder, Joseph J., and Cecil R. Phillips, *Project Management with CPM and PERT*, Reinhold Publishing Corporation, New York, 1964.

Moses, S., "CPM Opens a Pool Ten Months Early; West Hartford, Conn.," *American City*, vol. 79, pp. 78–79, December, 1964.

NASA PERT Handbook, National Aeronautics and Space Administration, Washington, D.C., July, 1961.

"Navy Extends PERT to Gauge Program Costs," *Missiles and Rockets*, vol. 10, p. 16, Jan. 15, 1962.

"Navy's PERT Way of Building Polaris Gives Industry a Potent New Management Weapon," *Purchasing Week*, Apr. 17, 1961.

Network Analysis, Lockheed Aircraft Corporation, March, 1961.

"Network Analysis; Progress of PERT," *Economist*, vol. 216, pp. 269+, July 17, 1965.

Neuwirth, S. I., and J. Zelnick, "Introduction to PERT," *Journal of Accountancy*, vol. 115, pp. 83–87, May, 1963.

"New Tools for Job Management," *Engineering News Record*, Jan. 28, 1961.

Norden, P., and O'Reilly, F., *Life Cycle Method of Project Planning and Control*, preliminary report paper on a study conducted at the Data System Division, IBM Product Development Laboratory, Poughkeepsie, N.Y.

"Origin of PERT, The," *The Controller,* December, 1962.

Page, J. C., and J. F. Stolle, "Space Age Technique to Launch New Products (PERT Program)," *Sales Management,* vol. 93, pp. 23–27+, July 3, 1964.

Paige, H. W., "How PERT/Cost Helps the General Manager," *Harvard Business Review,* vol. 41, pp. 87–95, November, 1963.

Pearlman, Jerome, "PERT: An Empirical Approach to Resources Planning," abstract of paper no. 5.1 presented at the 1961 International Convention of the Institute of Radio Engineers, Mar. 20, 1961, *Proceedings of the IRE,* March, 1961.

Pearlman, Jerry, AIA, "Presentation on PERT System at General Electric Light Military Electronics Division," General Electric Light Military Electronics Division, Utica, N.Y., July, 1960.

Perry, D. G., "Use of PERT in Systems Design," *NAA Bulletin,* vol. 45, August, 1964.

"PERT and CPM; New Planning Tools for Purchasing Management; Special Report," *Purchasing,* vol. 54, pp. 71–90, June 3, 1963.

"PERT: A Proven Project Planning Pill," *Office Administration,* October, 1962.

"PERT and You: Planning Control Communication," *Administrative, Management,* March, 1961.

PERT Application at IMED, IMED, General Electric, Utica, N.Y., March, 1961.

PERT: A Dynamic Project Planning and Control Method, Data Processing Division, IBM Corporation, October, 1961.

"PERT Is Extended to Control Project Costs," *Electronics,* Nov. 17, 1961.

"PERT: New Management Tool for Project Control," *Textile World,* vol. 113, pp. 129–131, November, 1963.

"PERT Is Pertinent to Polaris as a New Means to the End," *Western Aviation, Missiles and Space,* May, 1961.

"PERT: Pro and Con about This Technique," *Data Processing* (U.S.), October, 1961.

"PERT: A Recent Control Concept," *NAA Bulletin,* January, 1962.

"PERT Technique," *Data Processing for Management,* February, 1963.

Phelps, H. S., "PERT: What It Is and How It Works," *Supervisory Management,* vol. 7, pp. 20–25, December, 1962.

Phelps, H. S., "What Your Key People Should Know about PERT," *Management Review,* vol. 51, pp. 44–51, October, 1962.

"Planning, Implementation, and Appraisal through PERT," *Business Budgeting,* January, 1962.

"Pocket PERT (Pocket-sized PERT-O-Graph)," *Factory,* vol. 120, pp. 80–81, July, 1962.

Pocock, J. W., "PERT as an Analytical Aid for Program Planning: Its Payoff and Problems," *Operations Research,* vol. 10, pp. 893–903, November, 1962.

Program Evaluation and Review Technique, Operations Research, Inc., Los Angeles, 1961.

"Power for Peace," *Time Magazine,* Aug. 1, 1960.

"Progress Reporting in the Special Progress Office for the Fleet Ballistic Missile Program," *Navy Management Review,* April, 1959.

Rich, G., and T. Matye, *PERTING by E.A.M.,* Hughes Aircraft Compan, Culver City, Calif., Nov. 1, 1960.

Roman, Daniel D., "The PERT System: An Appraisal of Program Evaluation Review Technique," *Journal of Academy of Management,* April, 1962.

Sayer, J. S., J. E. Kelley, Jr., and Morgan R. Walker, "Presentation on PERT System," *Factory,* July, 1960.

Sando, F. A., "CPM: What Factors Determine Its Success?" *Architectural Record,* vol. 155, pp. 211–216, 202–204, April, May, 1964.

Schriever, Lt. Gen. Bernard A., "We Had Youth," *Armed Forces Management,* February, 1961.

"Second Look at PERT, A," *Lybrand Journal,* no. 3, vol. 43, 1962.

Shaffer, Louis R., J. B. Ritter, and W. L. Meyer, *Critical Path Method,* McGraw-Hill Book Company, New York, 1965.

"Shipfitting according to Computer (PERT)," *Economist,* vol. 208, p. 163, July 13, 1963.

"Shortcut for Project Planning; PERT/Cost Is Hottest New Tool in Space Age Research and Development," *Business Week,* pp. 104+, July 7, 1962.

Shultis, R. L., "Applying PERT to Standard Cost Revisions," *NAA Bulletin,* vol. 44, pp. 35–43, September, 1962.

Solomon, N. B., "Automated Methods in PERT Processing," *Computers and Automation,* vol. 14, pp. 18–22+, January, 1965.

"Space Age Scheduling Arrives in CPI," *Chemical Week,* Oct. 15, 1960.

Stilian, G. N., and others, *PERT: A New Management Planning and*

Control Technique, American Management Association, New York, 1963.

"Swedes Adopt PERT for Viggen Program," *Aviation Week,* vol. 82, pp. 257+, June 14, 1965.

"Teaching PERT Project Network Techniques," *Training Directors,* December, 1961.

Thompson, Van B., "PERT: Pro and Con about This Technique," *Data Processing* (U.S.), October, 1961.

"Time Scale Simplified CPS Diagrams," *Plant Administration & Engineering,* March, 1962.

"Tiny Computer Helps PERT Job (Called PERT-O-Graph)," *Business Week,* p. 116, Sept. 8, 1962.

"Use of PERT at Aeronautical Systems Division," *Journal of the Armed Forces Management Association,* February, 1962.

Usry, M. F., "PERT/Cost and the Capital Expenditure Control Program," *Journal of Accountancy,* vol. 115, pp. 83–86, March, 1963.

Van Krugel, E., "Introduction to CPM," *Architectural Record,* vol. 136, pp. 337+, September, 1964.

Van Slyke, R. M., "Monte Carlo Methods and the PERT Problem," *Operations Research,* vol. 11, pp. 839–860, September, 1963.

Villers, Raymond, "The Scheduling of Engineering Research," *Journal of Industrial Engineering,* November-December, 1959.

Wahl, R. P., Jr., "PERT Controls Budget Preparation (Fairfax County, Va.)," *Public Management,* vol. 46, pp. 29–33, February, 1964.

Waldron, A. James, *Fundamentals of Project Planning and Control,* A. James Waldron, West Haddonfield, N.J.

Wiest, J. D., "Some Properties of Schedules for Large Projects with Limited Resources (Critical Path Method, PERT, and Related Techniques)," *Operations Research,* vol. 12, pp. 395–418, May, 1964.

Square Roots (1-400)

1	1.00	41	6.40	81	9.00	121	11.00	161	12.69
2	1.41	42	6.48	82	9.06	122	11.05	162	12.73
3	1.73	43	6.56	83	9.11	123	11.09	163	12.77
4	2.00	44	6.63	84	9.17	124	11.14	164	12.81
5	2.24	45	6.71	85	9.22	125	11.18	165	12.85
6	2.45	46	6.78	86	9.27	126	11.23	166	12.88
7	2.65	47	6.86	87	9.33	127	11.27	167	12.92
8	2.83	48	6.93	88	9.38	128	11.31	168	12.96
9	3.00	49	7.00	89	9.43	129	11.36	169	13.00
10	3.16	50	7.07	90	9.49	130	11.40	170	13.04
11	3.32	51	7.14	91	9.54	131	11.45	171	13.08
12	3.46	52	7.21	92	9.59	132	11.49	172	13.11
13	3.61	53	7.28	93	9.64	133	11.53	173	13.15
14	3.74	54	7.35	94	9.70	134	11.58	174	13.19
15	3.87	55	7.42	95	9.75	135	11.62	175	13.23
16	4.00	56	7.48	96	9.80	136	11.66	176	13.27
17	4.12	57	7.55	97	9.85	137	11.70	177	13.30
18	4.24	58	7.62	98	9.90	138	11.74	178	13.34
19	4.36	59	7.68	99	9.95	139	11.79	179	13.38
20	4.47	60	7.75	100	10.00	140	11.83	180	13.42
21	4.58	61	7.81	101	10.05	141	11.87	181	13.45
22	4.69	62	7.87	102	10.10	142	11.92	182	13.49
23	4.80	63	7.94	103	10.15	143	11.96	183	13.53
24	4.90	64	8.00	104	10.20	144	12.00	184	13.56
25	5.00	65	8.06	105	10.25	145	12.04	185	13.60
26	5.10	66	8.12	106	10.30	146	12.08	186	13.64
27	5.20	67	8.19	107	10.34	147	12.12	187	13.67
28	5.29	68	8.25	108	10.39	148	12.17	188	13.71
29	5.39	69	8.31	109	10.44	149	12.21	189	13.75
30	5.48	70	8.37	110	10.49	150	12.25	190	13.78
31	5.57	71	8.43	111	10.54	151	12.29	191	13.82
32	5.66	72	8.49	112	10.58	152	12.33	192	13.86
33	5.74	73	8.54	113	10.63	153	12.37	193	13.89
34	5.83	74	8.60	114	10.68	154	12.41	194	13.93
35	5.92	75	8.66	115	10.72	155	12.45	195	13.96
36	6.00	76	8.72	116	10.77	156	12.49	196	14.00
37	6.08	77	8.77	117	10.82	157	12.53	197	14.04
38	6.16	78	8.83	118	10.86	158	12.57	198	14.07
39	6.25	79	8.89	119	10.91	159	12.61	199	14.11
40	6.32	80	8.94	120	10.95	160	12.65	200	14.14

201	14.18	241	15.52	281	16.76	321	17.92	361	19.00
202	14.21	242	15.56	282	16.79	322	17.94	362	19.03
203	14.25	243	15.59	283	16.82	323	17.97	363	19.05
204	14.28	244	15.62	284	16.85	324	18.00	364	19.08
205	14.32	245	15.65	285	16.88	325	18.03	365	19.11
206	14.35	246	15.68	286	16.91	326	18.06	366	19.13
207	14.39	247	15.72	287	16.94	327	18.08	367	19.16
208	14.42	248	15.75	288	16.97	328	18.11	368	19.18
209	14.46	249	15.78	289	17.00	329	18.14	369	19.21
210	14.49	250	15.81	290	17.03	330	18.17	370	19.24
211	14.53	251	15.84	291	17.06	331	18.19	371	19.26
212	14.56	252	15.87	292	17.09	332	18.22	372	19.29
213	14.59	253	15.91	293	17.12	333	18.25	373	19.31
214	14.63	254	15.94	294	17.15	334	18.28	374	19.34
215	14.66	255	15.97	295	17.18	335	18.30	375	19.36
216	14.70	256	16.00	296	17.20	336	18.33	376	19.39
217	14.73	257	16.03	297	17.23	337	18.36	377	19.42
218	14.76	258	16.06	298	17.26	338	18.38	378	19.44
219	14.80	259	16.09	299	17.29	339	18.41	379	19.47
220	14.83	260	16.12	300	17.32	340	18.44	380	19.49
221	14.87	261	16.16	301	17.35	341	18.47	381	19.52
222	14.90	262	16.19	302	17.38	342	18.49	382	19.54
223	14.93	263	16.22	303	17.41	343	18.52	383	19.57
224	14.97	264	16.25	304	17.44	344	18.55	384	19.60
225	15.00	265	16.28	305	17.46	345	18.57	385	19.62
226	15.03	266	16.31	306	17.49	346	18.60	386	19.65
227	15.07	267	16.34	307	17.52	347	18.63	387	19.67
228	15.10	268	16.37	308	17.55	348	18.65	388	19.70
229	15.13	269	16.40	309	17.58	349	18.68	389	19.72
230	15.17	270	16.43	310	17.61	350	18.71	390	19.75
231	15.20	271	16.46	311	17.64	351	18.74	391	19.77
232	15.23	272	16.49	312	17.66	352	18.76	392	19.80
233	15.26	273	16.52	313	17.69	353	18.79	393	19.82
234	15.30	274	16.55	314	17.72	354	18.81	394	19.85
235	15.33	275	16.58	315	17.75	355	18.84	395	19.87
236	15.36	276	16.61	316	17.78	356	18.87	396	19.90
237	15.39	277	16.64	317	17.80	357	18.89	397	19.92
238	15.43	278	16.67	318	17.83	358	18.92	398	19.95
239	15.46	279	16.70	319	17.86	359	18.95	399	19.98
240	15.49	280	16.73	320	17.89	360	18.97	400	20.00

Areas under the Curve

+ or −	.00	.01	.02	.03	.04	.05	.06	.07	.08	.09
0.0	.50000	.50399	.50798	.51197	.51595	.51994	.52392	.52790	.53188	.53586
0.1	.53983	.54380	.54776	.55172	.55567	.55962	.56356	.56749	.57142	.57535
0.2	.57926	.58317	.58706	.59095	.59483	.59871	.60257	.60642	.61026	.61409
0.3	.61791	.62172	.62552	.62930	.63307	.63683	.64058	.64431	.64803	.65173
0.4	.65542	.65910	.66276	.66640	.67003	.67364	.67724	.68082	.68439	.68793
0.5	.69146	.69497	.69847	.70194	.70540	.70884	.71226	.71566	.71904	.72240
0.6	.72575	.72907	.73237	.73536	.73891	.74215	.74537	.74857	.75175	.75490
0.7	.75804	.76115	.76424	.76730	.77035	.77337	.77637	.77935	.78230	.78524
0.8	.78814	.79103	.79389	.79673	.79955	.80234	.80511	.80785	.81057	.81327
0.9	.81594	.81859	.82121	.82381	.82639	.82894	.83147	.83398	.83646	.83891
1.0	.84134	.84375	.84614	.84849	.85083	.85314	.85543	.85769	.85993	.86214
1.1	.86433	.86650	.86864	.87076	.87286	.87493	.87698	.87900	.88100	.88298
1.2	.88493	.88686	.88877	.89065	.89251	.89435	.89617	.89796	.89973	.90147
1.3	.90320	.90490	.90658	.90824	.90988	.91149	.91309	.91466	.91621	.91774
1.4	.91924	.92073	.92220	.92364	.92507	.92647	.92785	.92922	.93056	.93189
1.5	.93319	.93448	.93574	.93699	.93822	.93943	.94062	.94179	.94295	.94408
1.6	.94520	.94630	.94738	.94845	.94950	.95053	.95154	.95254	.95352	.95449
1.7	.95543	.95637	.95728	.95818	.95907	.95994	.96080	.96164	.96246	.96327
1.8	.96407	.96485	.96562	.96638	.96712	.96784	.96856	.96926	.96995	.97062
1.9	.97128	.97193	.97257	.97320	.97381	.97441	.97500	.97558	.97615	.97670

z	.00	.01	.02	.03	.04	.05	.06	.07	.08	.09
2.0	.97725	.97784	.97831	.97882	.97932	.97982	.98030	.98077	.98124	.98169
2.1	.98214	.98257	.98300	.98341	.98382	.98422	.98461	.98500	.98537	.98574
2.2	.98610	.98645	.98679	.98713	.98745	.98778	.98809	.98840	.98870	.98899
2.3	.98928	.98956	.98983	.99010	.99036	.99061	.99086	.99111	.99134	.99158
2.4	.99180	.99202	.99224	.99245	.99266	.99286	.99305	.99324	.99343	.99361
2.5	.99379	.99396	.99413	.99430	.99446	.99461	.99477	.99492	.99506	.99520
2.6	.99534	.99547	.99560	.99573	.99585	.99598	.99609	.99621	.99632	.99643
2.7	.99653	.99664	.99674	.99683	.99693	.99702	.99711	.99720	.99728	.99736
2.8	.99744	.99752	.99760	.99767	.99774	.99781	.99788	.99795	.99801	.99807
2.9	.99813	.99819	.99825	.99831	.99836	.99841	.99846	.99851	.99856	.99861
3.0	.99865	.99869	.99874	.99878	.99882	.99886	.99889	.99893	.99896	.99900
3.1	.99903	.99906	.99910	.99913	.99916	.99918	.99921	.99924	.99926	.99929
3.2	.99931	.99934	.99936	.99938	.99940	.99942	.99944	.99946	.99948	.99950
3.3	.99952	.99953	.99955	.99957	.99958	.99960	.99961	.99962	.99964	.99965
3.4	.99966	.99968	.99969	.99970	.99971	.99972	.99973	.99974	.99975	.99976
3.5	.99977	.99978	.99978	.99979	.99980	.99981	.99981	.99982	.99983	.99983
3.6	.99984	.99985	.99985	.99986	.99986	.99987	.99987	.99988	.99988	.99989
3.7	.99989	.99990	.99990	.99990	.99991	.99991	.99992	.99992	.99992	.99992
3.8	.99993	.99993	.99993	.99994	.99994	.99994	.99994	.99995	.99995	.99995
3.9	.99995	.99995	.99996	.99996	.99996	.99996	.99996	.99996	.99997	.99997

Directions: To find the area under the curve between the left-hand end and any point, determine how many standard deviations that point is to the right of the average, then read the area directly from the body of the table. *Example:* The area under the curve from the left-hand end and a point 1.81 standard deviations to the right of the average is .96485 of the total area under the curve.

Index

Index

ABLE (Activity Balance Line Evaluation), 151
Activities, 10, 11
  rearrangement of, 88–92
  series-connected, 89
  series-parallel, 90
Activity Balance Line Evaluation (ABLE), 151
Areas under the curve, 171–173

Beta distribution, 33–40, 43
Booz, Allen & Hamilton, 2
BUWEPS PERT milestone system, 151

COMET (Computer Operated Management Evaluation Techniques), 150
Computer Operated Management Evaluation Techniques (COMET), 150
Computer outputs, 115–119
  departmental report, 118
  event number report, 115
  latest allowable date report, 118
  slack time report, 116
Computer Program for Scheduling Time and Distributing Cost (SKED), 156

Computers, use of, 112–119
Continuous probability distribution, 29–33
Contractual Requirements Recording, Analysis, and Management (CRAM), 149
Cost Planning and Appraisal (CPA), 151
Costs, 132
  direct, 132
  indirect, 132
  utility, 132
CPA (Cost Planning and Appraisal), 151
CPM (critical path method), 9
CRAM (Contractual Requirements Recording, Analysis, and Management system), 149
Crash time-and-cost estimates, 122
Critical path, 63, 64, 135
Critical path method (CPM), 120–145

Departmental report, 118
Direct costs, 132
Distribution, continuous probability, 29–33

Earliest expected date, 54–60

Event number report, 115
Events, 10, 11, 55

Gantt, H. L., 3, 11
Gantt charts, 3–5

HEPP (Hoffman Evaluation Program and Procedure), 151
Hoffman Evaluation Program and Procedure (HEPP), 151

ICON (Integrated Control), 152
IMPACT (Implementation, Planning, and Control Technique), 149
Implementation, Planning and Control Technique (IMPACT), 149
Indirect costs, 132
Integrated Control (ICON), 152
Interface events, 49

Latest allowable date, 64–71
Latest allowable date report, 118
Least Cost Estimating and Scheduling (LESS), 150
LESS (Least Cost Estimating and Scheduling technique), 150
Line of Balance (LOB), 148
LOB (Line of Balance), 148

Management Planning and Control System (MPACS), 152
Manpower Utilization, 148
Most likely time, 41, 93
Most optimistic time, 40, 93, 95
Most pessimistic time, 40, 93, 96
MPACS (Management Planning and Control System), 152

NASA PERT and Companion Cost, 152
Navy Special Projects Office, 2, 7
Network, 13
Network-beginning event, 15
Network-ending event, 15
Network replanning and adjustment, 78–92
Networking principles, 46–77
Networks, combining, 46–51
Nonrepetitive operations, 1, 2
Normal curve, 30
Normal time-and-cost estimates, 122

Operations, 1, 2
  nonrepetitive, 1, 2
  repetitive, 1, 2

PAAC (Program Analysis Adaptable Control Technique), 153
PAR (Project Audit Report), 154
PEP (Program Evaluation Procedure), 153
PERT (Program Evaluation and Review Technique), 3–5
  background of, 3–5
    Gantt charts, 3–5
    Polaris project, 7, 8
    transition steps, 5–7
  definitions, 2
  introduction, 1–9
PERT fundamentals, 10–18
PERT II, 154
PERT III, 154
PERT IV, 154
PERT/Cost, 153
PLANNET (Planning Network), 155
Planning Network (PLANNET), 155
Polaris project, 7, 8
PRISM (Program Reliability Information System for Management), 155
Probability concepts, 93–111

Program Analysis Adaptable Control Technique (PAAC), 153
Program Evaluation Procedure (PEP), 153
Program Reliability Information System for Management (PRISM), 155
Project Audit Report (PAR), 154

RAMPS (Resources Allocation and Multi-Project Scheduling), 155
Repetitive operations, 1, 2
Reports, computer prepared, 115–119
  departmental, 118
  event number, 115
  latest allowable date, 118
  slack time, 116
Resources, interchanging, 79–87
Resources Allocation and Multi-Project Scheduling (RAMPS), 155

SCANS (Scheduling and Control by Automated Network Systems), 155
Schedule Performance Evaluation and Review Technique (SPERT), 156
Scheduling and Control by Automated Network Systems (SCANS), 155
Series-connected activities, 15, 89
Series-parallel activities, 15, 90
SKED (Computer Program for Scheduling Time and Distributing Cost), 156
Slack time, 71–77, 109, 110
Slack time report, 116
SPERT (Schedule Performance Evaluation and Review Technique), 156

Square roots, 167–169
Standard deviation, 31, 96, 109, 111

Task Reporting and Current Evaluation (TRACE), 157
Taylor, F. W., 3
Technical specifications, relaxing of, 87, 88
The Operational PERT System (TOPS), 157
Time, 40, 41, 93, 95, 96
  most likely, 41, 93
  most optimistic, 40, 93, 95
  most pessimistic, 40, 93, 96
Time considerations, 25–45
Time-and-cost estimates, 122
  crash, 122
  normal, 122
TOES (Trade-Off Evaluation System), 156
TOPS (The Operational PERT System), 157
TRACE (Task Reporting and Current Evaluation), 157
Trade-Off Evaluation System (TOES), 156
Transition steps, 5–7

Utility costs, 132

Weapon Systems Programming and Control System (WSPACS), 157
WSPACS (Weapon Systems Programming and Control System), 157
Work breakdown schedule, 19–24

Zero-time activities, 14, 52–54, 60–62